全国现代学徒制工作专家指导委员会指导

Practical English of Cosmetology

美容实用英语

主　编　付明明

副主编　张　莹　龚丽萍　黄　晴

编　委　付明明（黑龙江中医药大学佳木斯学院）

　　　　张　莹（辽宁医药职业学院）

　　　　龚丽萍（中山市技师学院）

　　　　黄　晴（惠州卫生职业技术学院）

　　　　李艳微（黑龙江中医药大学佳木斯学院）

　　　　何丽琴（江西中医药高等专科学校）

　　　　马智利（珠海市卫生学校）

　　　　葛　露（辽宁医药职业学院）

　　　　张　娟（西安海棠职业学院）

复旦大學出版社

内容提要

本教材主要讲授美容实用英语，使学生掌握美容工作者日常的工作用语；会用英语表达专业的医美服务话术、美容皮肤护理话术及美容保健话术；帮助学生学习并掌握美容护理产品的英语表达方式，并能够用英语对客户进行介绍。培养学生能够熟练地运用美容英语与客户进行沟通和交流，并能够将所学知识灵活运用。

本教材作为现代学徒制美容教材，编写时以培养学生能力为本位，以岗位技能要求为目标，彻底打破原有课程体系，重新构建课程框架。本教材既适用于高职医学美容技术专业，又适用于中职美容美体、中医康复保健专业，也可面向美容导师、美容顾问、高级美容师、技术主管等岗位。

本套系列教材配有相关的课件、视频等，欢迎教师完整填写学校信息来函免费获取：xdxtzfudan@163.com。

总 序 FOREWORD

2019年1月以来,党中央、国务院先后印发《中国教育现代化2035》《加快推进教育现代化实施方案(2018—2022年)》《国家职业教育改革实施方案》等文件,为推动新时代职业教育现代化设定了目标图、路线图、施工图。现代学徒制体现了职业教育作为一种教育类型的重要特征,突出了产教融合、校企合作,实现了工学结合、知行合一,有力推动了职业教育和培训体系完善。

现代学徒制是借鉴发达国家职业教育经验,结合我国国情,自2014年起在职业教育领域开展的一项重大改革试点。试点5年以来,分批布局了558个现代学徒制试点,覆盖1 480多个专业点,9万余名学生学徒直接受益。现代学徒制坚持岗位成才、工学交替,有力推动了学校招生和企业招工相衔接,实现了校企育人"双重主体",学生学徒"双重身份",有力推动了学校和企业在资源、人员、技术、管理及人才培养成本等方面有机衔接,推动教育链、人才链与产业链、创新链同频共振。

广东省委省政府按照国家的统一部署,大力推行现代学徒制,省教育厅及省教育研究院积极推动广东特色现代学徒制。一是立法支持。2018年通过的《广东省职业教育条例》,确定了现代学徒制的地方法律地位。二是资金支持。2019年1月广东省政府办公厅印发《广东省职业教育"扩容、提质、强服务"三年行动计划(2019—2021年)的通知》,明确了"对开展学徒培养的企业,根据不同职业(工种)的培训成本,可按规定给予每生每年4 000~6 000元的培训补贴"。三是招生招工同步推进。开始职业教育现代学徒制招生单独渠道,鼓励学校、企业及学生签订协议或合同,明确学生学徒双重身份,着力推进教育制度与劳动制度深度融合。四是统一教学标准。研制出15个省级现代学徒制专业教学标准并在全省推广应用。

清远职业技术学院是教育部首批现代学徒制试点单位,在全国较早开展试点工作,并牵头联合企业共同研制了"现代学徒制医学美容技术专业教学标准和课程标准"。为固化现代学徒制研究和实践成果,在全国现代学徒制专家指导委员会和全国卫生职业教育教学指导委员会的支持指导下,广东省卫生职业教育协会和医学美容技术专业产教研联盟牵头,联合全国50多所相关院校或企业参与,共同开发了"全国现代学徒制医学美容技术专业'十三五'规划教材"。本套教材按照需求导向、校企合作的原则开发,与传统的学科体系教材相比,具有以下三个特征。

1. 对接标准。本套教材依据现代学徒制医学美容技术专业的课程标准编写,而这些课程标准有机对接了职业标准。职业标准是基于专业岗位群的职业能力分析,从专业能力和职业素养两个维度,形成了730条职业能力点。将工作领域的职业标准转化为学习领域的课程标准,重构医学美容技术专业的课程体系及课程内容。

2. 任务驱动。本套教材以美容岗位的工作过程为主线,以典型工作任务驱动知识和技能的学习,让学生在做中学,在"会做"的同时,用心领悟"为什么做",应具备"哪些职业素养",教材结构及内容符合技术技能人才培养的基本要求。

3. 多元受众。本套教材由校企专家共同开发,教材内容既包含医学美容技术专业的理论知识,也涵盖医学美容岗位的技术技能;既能满足学历教育需求,也能满足职业培训需求。教材可供职业院校和行业企业多方开展教育教学使用。

现代学徒制医学美容技术专业系列教材的开发对于当前职业教育改革具有重要意义,它不仅是现代学徒制人才培养模式改革成果的重要形式之一,更是对职业教育现实需求的重要回应。作为全国现代学徒制首批试点专业所形成的这套教材,其开发路径与方法能为其他现代学徒制专业提供借鉴,起到抛砖引玉的作用。

全国现代学徒制工作专家指导委员会主任委员

广东建设职业技术学院院长

博士,教授

2019年3月

前言 PREFACE

在全国现代学徒制工作专家指导委员会和全国卫生职业教育教学指导委员会的支持指导下,广东省卫生职业教育协会和医学美容技术专业产教研联盟牵头,联合全国50多所相关院校和企业参与,共同开发了"全国现代学徒制医学美容技术专业'十三五'规划教材"。《美容实用英语》是本套教材之一。

美容实用英语在美容行业与国际接轨日趋紧密的情况下,其助推美容服务国际化的作用逐渐显现,掌握美容实用英语已经成为美容师的基本工作任务之一。依据《现代学徒制专业教学标准和课程标准:医学美容技术专业》中的美容实用英语课程标准(详见附录),基于美容师岗位的典型工作任务,我们编写了《美容实用英语》这本教材。本教材以英语实践技能操作为主导,突出基本日常实用英语的技能训练,特别是美容英语口语方面的灵活运用,让学生学习后,既能够按照美容服务的流程用英语使用标准话术,又能够根据实际情况与客户展开沟通,了解客户的所需,提供科学的指导方案,等等。《美容实用英语》从用英语进行服务流程演练,到产品说明与介绍,再到售后服务与人文关怀,为美容师顺利进入岗位工作打下基础,最大限度地满足美容企业的用人需求。本教材学习任务安排由浅入深逐步递进,适用于医学美容技术专业和中医美容专业,以及各类美容职业教育培训。

本教材的编写队伍来自全国各地职业院校。作为一本现代学徒制教材,本教材在编写过程中还得到了多家企业的大力支持。他们慷慨提供了各种指导和帮助,这才使得本教材能够顺利完成编写并真正与企业用人充分结合。在此对香港雅姬乐集团有限公司、长春市银河化妆品有限公司、美丽田园医疗健康产业有限公司、海南红瑞医疗美容投资管理有限公司、深圳市医美影视文化传媒有限公司以及佛山市金谷美妍健康咨询服务有限公司一并表示衷心的感谢!

由于时间紧迫,作者水平有限,书中难免出现疏漏和缺憾,恳请教材使用者批评指正,以帮助我们再版时改进。

编者

2019年3月

Contents

Part I — Daily Courtesy Language in Beauty Salon

Unit 1　Greetings 1-3
　　Task 1　Beauty Salon Greetings and Salon Guidance Services 1-3
　　Task 2　Understanding Needs and Registration 1-6

Unit 2　Telephone Invitation and Post-service Interview 1-10
　　Task 1　Activity Invitation 1-10
　　Task 2　Post-service Interview 1-14

Part II — Communication Skills in Cosmetology Domain

Unit 1　Communication Skills in Medical Beauty Service 2-3
　　Task 1　Customer Consultation Service 2-3
　　Task 2　Preoperative Communication 2-6
　　Task 3　Postoperative Precautions 2-9
　　Task 4　Postoperative Follow-up 2-12

Unit 2　Beauty Care and Skin Care 2-16
　　Task 1　Introduction of Service Process 2-16
　　Task 2　Preparation for Professional Facial Care 2-20
　　Task 3　Standard Process of Facial Care 2-23
　　Task 4　Home Treatment Advice 2-27

Unit 3　Beauty Care and Body Care 2-30
　　Task 1　Health Information of Customers 2-30
　　Task 2　Programs Introduction 2-34
　　Task 3　Health Care Service Process 2-37
　　Task 4　Health Care Tips 2-40

Part III Introduction of Cosmetics

Unit 1　Facial Care Products ·· 3-3
　　Task 1　Facial Cleanser ··· 3-3
　　Task 2　Exfoliating Gel ·· 3-6
　　Task 3　Toner ·· 3-10
　　Task 4　Eye Cream ··· 3-12
　　Task 5　Moisturizing Cream ··· 3-16
　　Task 6　Sunscreen ·· 3-19

Unit 2　Body Care Products ··· 3-24
　　Task 1　Carrier Oil & Essential Oil ···································· 3-24
　　Task 2　Slimming Cream ··· 3-28
　　Task 3　Hair Removal Spray Foam ···································· 3-32

Unit 3　Hair Care Products ·· 3-36
　　Task 1　Shampoo ··· 3-36
　　Task 2　Hair Mask ·· 3-39
　　Task 3　Hair Dye ·· 3-43
　　Task 4　Hair Tonic ·· 3-46

Part IV Practical Simulation Training

　　Task 1　Full Process Simulation Training for Medical Beauty Customer Service ··· 4-3
　　Task 2　Full Process Simulation Training for Beauty Skin Care Customer Service ··· 4-8

参考文献 ·· 1

附录　课程标准 ··· 2

Part I Daily Courtesy Language in Beauty Salon

Unit 1

Greetings

Learning Objectives

1. Learn to greet customers in English and give them guidance in beauty salon.
2. Be able to clearly understand the needs of customers and skillfully help customers to register.

 Beauty Salon Greetings and Salon Guidance Services

 Lead-in

This is the first time for a customer coming into the beauty salon; one staff member of the salon begins to communicate with the new customer.

Thinking and Talking

1. What are the appropriate standard greeting languages when the beautician guide

customers?

2. What kinds of service consciousness should we have?

New Words

customer /ˈkʌstəmə(r)/	n.	顾客,客户;主顾
beautician /bjuːˈtɪʃn/	n.	美容师,美容专家
service /ˈsɜːvɪs/	n.	服役;服务,服侍
consultant /kənˈsʌltənt/	n.	顾问
skin /skɪn/	n.	皮,皮肤;毛皮
program /ˈprəʊɡræm/	n.	项目;计划,安排
care /keə(r)/	n.	护理;照顾
bathrobe /ˈbɑːθrəʊb/	n.	浴衣,浴袍
disinfect /ˌdɪsɪnˈfekt/	v.	除去(感染),给……消毒
guide /ɡaɪd/	v.	引路;指导

Phrases and Expressions

have some tea	喝茶
this way, please	这边请
batching room	配料间
skin analysis apparatus	皮肤分析仪

Intensive Reading

Text

(It's the first time for Ms. Wang coming into the beauty salon.)

A: Beautician　B: Ms. Wang

A: Hello, Ms. Wang, please come in.
B: Hello.
A: Oh, Ms. Wang. What can I do for you?
B: I'd like to know the services of your salon.
A: OK, give me a moment. Please, have some tea first.
B: Give me a cup of flower tea, please.
A: Enjoy it first, and then I'll show you around our salon.

B: OK.

A: Shall we visit the salon now? This way, please, Ms. Wang.

B: OK.

A: The first room is our consultant room. As a new customer, we will first provide you with our skin analysis service. According to your skin condition, we will provide a high-quality beauty treatment. The second room is our batching room for preparing materials.

B: How about the third room?

A: The third room is our beauty care room. There are sterilized bedding, bathrobes and towels in every beauty care room. There is also a locker for you to temporarily store your belongings.

B: This room looks very luxurious.

A: Yes, it is for VIP customers.

B: OK, well, I want to learn more about the skin analysis apparatus.

A: OK, wait a minute, please. I'll ask our professional beauty consultant to help you.

Scanning Audio

1. Fill in the blanks with the words given below, changing the form if necessary.

| customer batching skin instrument locker disinfect bathrobe guide |

(1) The _____ gives some suggestions on the environment of this beauty salon.

(2) The _____ room is used for preparing customers' materials.

(3) The _____ is used for temporary storage of customers' belongings.

(4) The _____ can be used directly after bathing.

(5) You can rest assured that all indoor articles have been _____.

2. Match the Chinese phrases with right English expressions.

电话预约   beauty salon
皮肤检测   skin analysis
面部护理   facial care
身体护理   body care
美容会所   telephone reservation

3. Arrange the following steps in the correct order from start to finish.

(1) greeting customers (2) sending out customers (3) guiding in salon (4) helping with the registration

The correct order is: _____.

4. Translate the following sentences into English.
   (1) 很荣幸为您服务,欢迎您下次光临!
   (2) 请问您需要喝玫瑰花茶、咖啡还是白开水?
   (3) 请您检查贵重物品是否齐全。
   (4) 这边请。这是护理房间。
   (5) 希望您能满意我们的服务!

Critical Thinking and Group Peer-assessment

1. Change the customer types or demands, redesigning the conversation.
2. Each group tries to display the dialogue according to the text.
3. After one group finishes the dialogue, others make comments.

Knowledge Links

Standing Posture

1. The heels are close to each other. The feet are separated into "V" shape (male) or "T" shape (female). The opening angle is $45°\sim60°$. The center of body weight falls in the middle of two feet.
2. Legs erect, knees together.
3. Draw in the abdomen and hip.
4. Stand upright and straight back.
5. Flat shoulders, relax and sink.
6. Shoulders droop naturally, fingers bend naturally.
7. Head forward, neck straight, jaw slightly retracted, eyes straight ahead.

 Understanding Needs and Registration

This is the first time for a customer coming into the beauty salon, a beautician is trying to understand the needs of the customer and helping her to register.

Thinking and Talking

1. How to guide customer to register?
2. What is the important information of customers when registering?

 ## New Words

acne /ˈækni/	n. 痤疮，粉刺
product /ˈprɒdʌkt/	n. 产品；作品
occupation /ˌɒkjʊˈpeɪʃn/	n. 职业，工作
file /faɪl/	n. 文件夹；卷宗
information /ˌɪnfəˈmeɪʃn/	n. 消息；信息
record /ˈrekɔːd/	n. 记录，记载；档案
/rɪˈkɔːd/	v. 记录，记载；标明；录音
check /tʃek/	v. 检查，核对
satisfy /ˈsætɪsfaɪ/	v. 使满意，满足

 ## Phrases and Expressions

internal conditioning	内部调理
external care	外部护理
obvious improvement	明显改善

 ## Intensive Reading

Text

(After Ms. Wang finishes the guided tour in the beauty salon, the beauty consultant tries to learn more about Ms. Wang.)

A: Beauty consultant B: Ms. Wang

A: Hello, Ms. Wang, my name is Hongmei. Do you like our beauty salon?

B: I like it very much.

A: What kind of service can I do for you? Facial care, body care or SPA?

B: I mainly want to have an acne facial treatment.

A: Our program combines internal conditioning with external care.

B: How long will it take effect?

A: After one week, the acne-damaged skin will be repaired, and you'll see an obvious improvement.

B: What is the price?

A: There is a promotion in our salon. According to your skin, we could also add some other products. Compared with the original price of 6,888 RMB, it is only 3,888 RMB now.

B: OK, this one, please.

A: Good, I will make a beauty registration card for you.

B: OK.

A: May I have your name, birthday, occupation and telephone number?

B: Wang Lin, born on June 28, 1994, company employee and my mobile phone number is 139********.

A: OK, thank you, Ms. Wang. I have set up a customer file for you. It contains your identity information, basic information of skin analysis and the information about your products. We will make a detailed record after service every time. Please check up the information again.

B: OK, no problem.

1. Fill in the blanks with the words given below, changing the form if necessary.

   | satisfy care sign activity occupation file check cash |

   (1) Here is your bill. Would you like to pay in _____ or by credit card?
   (2) What is your _____, teacher or doctor?
   (3) We'll make a customer _____ to collect your information.
   (4) Registration is over; please _____ up the information.
   (5) If there is nothing wrong, could you _____ your name, please?

2. Match the Chinese phrases with right English expressions.

   基本信息   eating habits
   登记注册   registration
   过度劳累   essential information
   兴趣爱好   hobbies and interests
   饮食习惯   overworked

3. Arrange the following steps in correct order.
   (1) understanding customer needs (2) registration and payment (3) providing goods and plans (4) setting up customer files
   The correct order is: _____.

4. Translate the following sentences into English.
   (1) 请您签一下名字好吗?
   (2) 这是您的账单,请核对账单上的详细内容。
   (3) 您目前对于皮肤的状况满意吗?
   (4) 请问您是现在付款吗? 刷卡还是现金?
   (5) 您皮肤以往有过过敏的情况吗?

Critical Thinking and Group Peer-assessment

1. Change the customer types or demands, redesigning the conversation.
2. Each group tries to display the dialogue according to the text.
3. After one group finishes the dialogue, others make comments.

Knowledge Links

Exercise Most Days of the Week

Findings from a few studies suggest that moderate exercise can improve circulation and boost the immune system. This, in turn, may give the skin a more youthful appearance.

https://www.aad.org/public/skin-hair-nails/anti-aging-skin-care/causes-of-aging-skin

Unit 2

Telephone Invitation and Post-service Interview

Learning Objectives

1. Learn to use daily courtesy language to invite customers and do post-service interview through telephone.
2. Be able to clearly introduce the promotion activities of the beauty salon.

Task 1 Activity Invitation

Lead-in

There is an anniversary celebration in the beauty salon. One beauty consultant of the salon is calling a customer and introducing the details.

Thinking and Talking

1. How to introduce yourself when you call customers for the first time?

2. What elements should we pay attention to when we call customers?

New Words

anniversary /ˌænɪˈvɜːsərɪ/	n.	周年纪念日
	a.	周年的
discount /ˈdɪskaʊnt/	v.	打折；减价
consider /kənˈsɪdə(r)/	v.	考虑；细想
dry /draɪ/	a.	干的；干燥的
aging /ˈeɪdʒɪŋ/	n.	老化；老龄化
deadline /ˈdedlaɪn/	n.	最后期限；截止时间
weekend /ˌwiːkˈend/	n.	周末
accept /əkˈsept/	v.	接受；承认
massage /məˈsɑːʒ/	n.	按摩，推拿

Phrases and Expressions

promotion price	活动价格
best seller	畅销品
fine lines	细纹，细线
hurry up	抓紧时间

Intensive Reading

Text

(The beauty salon is having an anniversary celebration, so one beauty consultant of the salon is calling Ms. Zhao.)

A: Beauty consultant   B: Ms. Zhao

A: Hello, Ms. Zhao, this is the beauty salon. I'm Xiaoxia, a consultant of the beauty salon. We are having an anniversary celebration now. Do you want to learn something about it?

B: Oh, yes. I want to know about discount programs on facial care.

A: What specific aspects do you want to know? For example, moisturizing, acne, freckles, or anti-aging?

B: Anti-aging products.

A: You may consider trying our anti-aging suite. The original price is 7,888 RMB and the promotion price is only 4,888 RMB.

B: Still sounds a little expensive.

A: This is just one of the discounted products, but the best seller.

B: Is there anything cheaper?

A: Anti-aging is a difficult one in beauty programs, so the price is more expensive.

B: Recently, I've got fine lines on my face and it sometimes feel dry and stinging.

A: It's winter now, Ms. Zhao. It's probably that hydropenia led to the pain and dryness.

B: How do I deal with the fine lines?

A: If the fine lines are caused by dryness, we can use moisturizing products. If they are caused by aging, we should use anti-aging products, so come in to our beauty salon.

B: All right. I will go have a look when I have time.

A: You'd better hurry up. Our special offers are available until the end of this week.

B: OK, see you on Saturday then.

Scanning Audio

1. Fill in the blanks with the words given below, changing the form if necessary.

> discount   series   best seller   weekend   promotion
> condition   deadline   accept   wrinkle

(1) Can you _____ this price?

(2) The usual price of this product is 2,000 RMB, and the _____ price is 800 RMB.

(3) This is a _____, and sold out as soon as it arrived.

(4) The _____ of this activity is weekend, and then it will restore the original price.

(5) I often wear masks and my skin _____ is not bad.

2. Match the Chinese phrases with right English expressions.

促销   spicy food
有效日期   on promotion
生产日期   effective date
油炸食品   production date
辛辣食物   fried food

3. Arrange the following steps in correct order.
   (1) spread the massage cream on the face   (2) massage the cheeks   (3) massage the eye area   (4) massage the forehead gently   (5) massage the mouth and nose area
   The correct order is: _____.

4. Translate the following sentences into English.
   (1) 您需不需要了解一下呢?
   (2) 您具体想要了解哪个方面呢?
   (3) 这只是众多优惠产品之一。
   (4) 有过敏时您都做过什么处理?
   (5) 您选择产品的时候,对产品的价格有什么样的要求?

Critical Thinking and Group Peer-assessment

1. Change the customer types or demands, redesigning the conversation.
2. Each group tries to display the dialogue according to the text.
3. After one group finishes the dialogue, others make comments.

Knowledge Links

Chinese Important Festivals

New Year's Day	元旦
Labor Day	劳动节
Women's Day	女人节
Teachers' Day	教师节
National Day	国庆节
Spring Festival	春节
Lantern Festival	元宵节
Dragon Boat Festival	端午节
Double Seventh Festival	七夕节
Mid-Autumn Festival (Moon Festival)	中秋节
Mother's Day	母亲节

Task 2 Post-service Interview

Lead-in

One beauty consultant of the beauty salon is making a phone call to a customer for post-service interview and trying to learn whether the customer is satisfied with the beauty salon and the service, then providing useful advice.

Thinking and Talking

1. What kinds of issues should we pay attention to when doing customer post-service interview?
2. How to deal with customer dissatisfaction?

New Words

conscientious	/ˌkɒnʃɪˈenʃəs/	a. 认真负责的
comfortable	/ˈkʌmftəbl/	a. 舒适的;安逸的
sweetmeat	/ˈswiːtmiːt/	n. 甜食;糖果;果脯
cooperate	/kəʊˈɒpəreɪt/	v. 合作,配合
persistent	/pəˈsɪstənt/	a. 坚持不懈的
perfect	/ˈpɜːfɪkt/	a. 完美的;正确
advance	/ədˈvɑːns/	a. 预先的

Phrases and Expressions

send WeChat	发微信
arrange time	安排时间
make an appointment	预约

Part I Daily Courtesy Language in Beauty Salon 1-15

 Intensive Reading

Text

(One beauty consultant of the beauty salon makes a
phone call to Ms. Wang for post-service interview.)

A: Beauty consultant   B: Ms. Wang

A: Hello, Ms. Wang, this is Xiaoxia, a beauty consultant at the beauty salon. May I have a post-service interview with you?

B: OK, no problem.

A: How do you like the environment and your beautician at the beauty salon?

B: The beautician is very conscientious and the environment is also comfortable.

A: How about our products?

B: There are no obvious effects so far. There is still severe acne on my face.

A: How about your sleep these days?

B: I go to bed at 12 p.m. and get up at 7 a.m.

A: What kind of food do you like? Spicy, sweet or fried food?

B: I like spicy food, not fried food, but I like sweetmeat very much.

A: Ms. Wang, in fact, the acne on your face has an important relationship with your work, rest and diet. Fundamentally, if you want to get rid of acne, you should cooperate with our beauty salon in terms of living habits.

B: What should I do?

A: I suggest that you should go to bed around 10 o'clock every night, eat light food, drink more water and eat more fruit.

B: OK, I'll try my best.

A: If you want to get rid of acne, you must persist. You look so beautiful! If there is no acne, you'll be more perfect!

B: OK, thank you. I'll be persistent on it.

A: If you have any question, please feel free to call me or send a WeChat.

B: OK, I will consult you whenever I have any questions.

A: It has been three days since your last skin treatment at our beauty salon. You should consider having another one.

B: OK. I'll make time tomorrow afternoon. I'll call you in advance to make an appointment.

A: OK. See you tomorrow.

Scanning Audio

Practice

1. Fill in the blanks with the words given below, changing the form if necessary.

conscientious favourable sweetmeat cooperate persist perfect daily WeChat

   (1) Do you satisfy with my service? Can you give me a _____ reviews?

   (2) You need to _____ for a long time to see the effects.

   (3) Nice to meet you. Could I have your _____ so that we could communicate with each other more easily?

   (4) The beauticians' service is very _____ in this beauty salon, so there is no complaint from customers.

   (5) I like eating _____, but eating too much may be fat, even bad for my skin.

2. Match the Chinese phrases with right English expressions.

   宣传单　　　　party
   庆祝　　　　　balanced diet
   聚会　　　　　leaflet
   诊断　　　　　diagnosis
   均衡饮食　　　celebration

3. Arrange the following festivals in correct order.

   (1) Women's Day (2) National Day (3) Mid-Autumn Festival (4) International Labor Day (5) Spring Festival

   The correct order is: _____.

4. Translate the following sentences into English.

   (1) 能耽误您几分钟时间做一个回访吗?

   (2) 您对环境和您的美容师还满意吧?

   (3) 您最近睡眠状况如何?

   (4) 您平时都喜欢吃什么口味的食物?

   (5) 您有什么疑问可以随时打电话或者给我发微信。

Critical Thinking and Group Peer-assessment

1. Change the customer types or demands, redesigning the conversation.
2. Each group tries to display the dialogue according to the text.
3. After one group finishes the dialogue, others make comments.

Knowledge Links

Gray Skin

Has your skin lost its' rosy, pink color? This may be normal when it comes to aging (and you can always fight aging with Life Cell All-In-One Anti-Aging Treatment to give you your glow back), but it can also mean that you may have kidney problems.

http://www.lifecellskin.com/news/are-you-listening-to-your-body/

Discussion Questions

1. What details should we pay attention to in the process of communicating with customers?
2. How to deal with customers' complaints?

Part II
Communication Skills in Cosmetology Domain

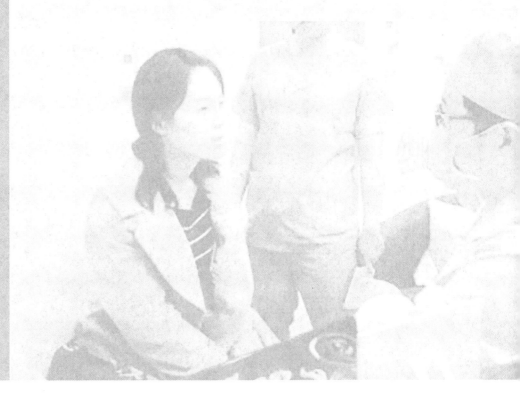

Unit 1

Communication Skills in Medical Beauty Service

Learning Objectives

1. Learn to provide medical beauty service for customers in English and give them in-store guidance.
2. Be able to support customers with preoperative guidance, postoperative precautions and postoperative follow-up.

Task 1 Customer Consultation Service

Lead-in

Ms. Li comes to the cosmetic and plastic surgery institution, the staff gives her thoughtful guidance and introduction of rhinoplasty surgery.

Thinking and Talking

1. How to guide the customers to register?
2. How to introduce rhinoplasty surgery?

 New Words

recover /rɪˈkʌvə/	v.	恢复
lesion /ˈliːʒn/	n.	病灶
limb /lɪm/	n.	肢,臂
organ /ˈɔːgən/	n.	器官
tissue /ˈtɪʃuː/	n.	组织

 Phrases and Expressions

rhinoplasty surgery	隆鼻手术
injection rhinoplasty	注射隆鼻
postoperative care	术后护理
injection area	注射区域

 Intensive Reading

Text

(Ms. Li comes to the cosmetic and plastic surgery institution, consulting an adviser about the rhinoplasty surgery.)

A: Consultant   B: Ms. Li

A: Good morning, Ms. Li. Can I help you?

B: Hello, I want to consult with you about the rhinoplasty surgery. Which kind of surgery is safer and better?

A: OK, Ms. Li. Injection rhinoplasty is better and safer.

B: How long is recovery after surgery?

A: It varies from person to person. Some people may take 1 week to recover after the surgery, while others may take 2~3 weeks.

B: What should I pay attention to before the surgery?

A: You should be sure there are no lesions on your face before the surgery.

B: How do I perform postoperative care?

Part II Communication Skills in Cosmetology Domain

A: Please avoid touching the injection area within 6 hours after the rhinoplasty surgery.
B: What about the price?
A: Are you a VIP member of our cosmetic and plastic surgery institution? In our recent promotion, we offered our members a discount of 20%. You can also enjoy this discount if you buy products in our salon.
B: I'm a VIP member, and please make an appointment for the rhinoplasty surgery for me.
A: Thanks for your trust. I will make the appointment for you right now.

Scanning Audio

1. **Fill in the blanks with the words given below, changing the form if necessary.**

   > prosthesis(假体)   ear cartilages(软骨)   mini-plastic surgery   facial plastic
   > nasal surgery   epicanthal plasty   mat chin(垫下巴)   lip enhancement(丰唇)
   > rhinoplasty recovery(隆鼻修复)   nasal aesthetic(鼻部美学)

   （1）_____ is mainly used for people whose lips are too thin.
   （2）The key step after the rhinoplasty surgery is _____.
   （3）_____ aims to beautify the chin.
   （4）_____ focuses on people's nose beauty.
   （5）_____ is a kind of medical device that replaces people's limbs, organs or tissues.

2. **Match the Chinese phrases with right English expressions.**

   微创技术   rhinoplasty surgery
   注射隆鼻   minimally invasive technology
   手术痕迹   nasal aesthetic
   隆鼻手术   injection rhinoplasty
   鼻部美学   surgery trace

3. **Arrange the following steps in the correct order.**
   （1）introducing rhinoplasty surgery   （2）understanding customers' needs   （3）sending out customers   （4）applying for VIP   （5）greeting customers
   The correct order is: _____.

4. **Translate the following sentences into English.**
   （1）隆鼻手术是一种比较普遍的医美微型手术。
   （2）隆鼻手术可以提高面部立体感。
   （3）隆鼻手术必须年满18周岁。

2-6　美容实用英语

（4）隆鼻手术前需要注意什么？
（5）隆鼻手术后怎样进行护理？

 Critical Thinking and Group Peer-assessment

1. Change the customer types or demands, redesigning the conversation.
2. Each group tries to display the dialogue according to the text.
3. After one group finishes the dialogue, others make comments.

Knowledge Links

Nose plastic must ensure the harmony between the nose and the face. For example, a long face with nostril（鼻孔）slightly up nose, can make the face appear shorter, and the correction of the upturned nose（朝天鼻）will break the face balance. Each person has unique facial bones, so it's important to choose a suitable personalized program for yourself.

 Task 2　Preoperative Communication

 Lead-in

Ms. Li comes to the cosmetic and plastic surgery institution, and the staff gives her thoughtful guidance and introduction of the double eyelid surgery.

Thinking and Talking

1. How to get customers' background information?
2. The consultant introduces the communication skills required for the plastic surgery.

 New Words

psychological /ˌsaɪkəˈlɒdʒɪkl/	a. 心理的；精神上的
hypertension /ˌhaɪpəˈtenʃn/	n. 高血压

diabetes /ˌdaɪəˈbiːtiːz/	n. 糖尿病
anticoagulant /ˌæntɪkəʊˈæɡjʊlənt/	n. 抗凝药
menstruate /ˈmenstrʊeɪt/	v. 行经
pregnant /ˈpreɡnənt/	a. 怀孕的
multiple /ˈmʌltɪpl/	a. 多重的；多样的

Phrases and Expressions

history of heart disease	心脏病病史
invigorating qi and promoting blood circulation	补气活血
blood test	验血
coagulation examination	凝血检查
preoperative immune	术前免疫

Intensive Reading

Text

(Ms. Li comes to the cosmetic and plastic surgery institution, for consultation about the precautions of double eyelid surgery.)

A: Consultant   B: Ms. Li

There are two kinds of preparation before microsurgery: one is psychological preparation, meaning you can't have unrealistic expections; the other preparation is preoperative examinations by the hospital, especially for elderly people.

A: Hello, Ms. Li. What can I do for you?

B: Hello, I'd like to know of the preparations before the double eyelid surgery.

A: Do you suffer from hypertension, diabetes or heart disease?

B: No.

A: Do you take any oral anticoagulant, vitamin E, Traditional Chinese Medicine for invigorating qi and promoting blood circulation? If yes, you need to stop taking the medicine more than 2 weeks before the surgery.

B: No.

A: We also need to know if you are menstruating or pregnant, and if you have a fever or a cold.

B: No.

A: You also need to complete preoperative examinations, including blood tests, a coagulation examination, preoperative immune 8 items, an electrocardiogram etc.

B: OK.

A: Finally, it is suggested you take a bath the night before the surgery and put on clean and comfortable clothes. No smoking or drinking for at least 1 week before the surgery, and at least 2 weeks for some special surgeries.

B: OK, thank you very much.

Scanning Audio

1. Fill in the blanks with the words given below, changing the form if necessary.

---

embedded line   parallel type   fan type   single eyelid   suture(缝合)   epicanthi(内眦赘皮)   eyebrow excision(切眉术)

---

(1) In double eyelid types, _____ is that the double eyelid is basically parallel to the palpebral margin(睑弦).

(2) _____ is also known as buried line, a line sewn into the upper eyelid to join the epidermis(表皮) and fascia(筋膜) of the eyelid.

(3) _____ is the removal of part of the brow and surrounding skin to reshape the brow.

(4) Chinese residents' eyelids can be divided into _____, double eyelid and multiple double eyelids.

(5) After double eyelid surgery, _____ can appear.

2. Match the Chinese phrases with right English expressions.

双眼皮　　　　　embedded line
埋线　　　　　　inner corner of the eye
全切　　　　　　upper eyelid
内眼角　　　　　double eyelid
上睑　　　　　　full cut

3. Arrange the following steps in the correct order.

(1) introducing double eyelid surgery   (2) no drinking after surgery   (3) healthy requirements before surgery   (4) welcoming customers   (5) guidance in salon

The correct order is: _____.

Part II Communication Skills in Cosmetology Domain   2-9

4. Translate the following sentences into English.
   (1) 双眼皮手术包括埋线、韩式三点和全切。
   (2) 手术前需要做什么检查?
   (3) 请问您有无高血压、糖尿病、心脏病病史?
   (4) 您是否在月经期或怀孕期?您是否在发烧、感冒的状态?
   (5) 非常感谢您的术前指导。我会尽快确定手术时间,打电话给前台。

 Critical Thinking and Group Peer-assessment

1. Change the customer types or demands, redesigning the conversation.
2. Each group tries to display the dialogue according to the text.
3. After one group finishes the dialogue, others make comments.

 Knowledge Links

　　Double eyelid surgery began in ancient Greeks and Romans. Anatomically, an eyelid is consisted of skin, soft tissue, fat and muscle for eye opening and tarsal plate(睑板). The eyelid surgeries are called blepharoplasties (眼睑整容术) and are performed either for medical reasons or to alter one's facial appearance.

Task 3 Postoperative Precautions

 Lead-in

　　After Ms. Li's double eyelid surgery, the postoperative precautions are suggested by the consultant.

Thinking and Talking
   1. How to grasp the surgery state after double eyelid surgery?
   2. The consultant introduces double eyelid surgery precautions to customers.

New Words

cosmetics /kɒzˈmetɪks/	n. 化妆品
sauna /ˈsɔːnə/	n. 桑拿
induration /ˌɪndjʊˈreɪʃən/	n. 硬结
embossment /ɪmˈbɒsmənt/	n. 凸起
headache /ˈhedeɪk/	n. 头痛
tingling /ˈtɪŋglɪŋ/	n. 麻刺感
nausea /ˈnɔːzɪə/	n. 恶心
metabolism /məˈtæbəlɪzəm/	n. 新陈代谢

Phrases and Expressions

red and swollen	红肿
surgical site	手术部位
stationary state	静止状态
light diet	清淡饮食
local muscle weakness	局部肌无力
strenuous exercise	剧烈运动

Text

Introducing postoperative precautions

According to individual differences during surgery, local reactions after surgery and notice.

1. Ice compress can reduce redness and swelling.
2. Do not touch the surgical site within 6 hours after surgery, and apply cosmetics externally 24 hours later.
3. Try to keep surgical site stationary within 48 hours after surgery.
4. Do not drink, smoke, sauna or other activities after surgery.
5. Induration and embossment may appear after surgery, and then disappear 3 months later.

6. Wrinkle removal needle can improve unnatural facial expression after surgery.
7. Light diet is needed after surgery; avoid raw, cold and spicy food.
8. A few people may suffer from headache, local muscle weakness, skin tightness, tingling, nausea etc., which can be relieved after symptomatic treatment.
9. Avoid strenuous exercise within 3 weeks after surgery.

 Practice

1. Fill in the blanks with the words given below, changing the form if necessary.

> double eyelid surgery　inflating　induration　rhytidectomy(除皱术)　suture material(缝合材料)　face-lift needle(瘦脸针)　canthus secretion(眼角分泌物)

(1) _____ can reduce wrinkle and improve dermatolysis(面部皮肤松弛).
(2) After double eyelid surgery, a kind of thick pale yellow liquid, _____ may appear.
(3) _____ is drug therapy of thinning face, certain risks exist.
(4) _____ and embossment after surgery, they can disappear after hot compress.
(5) Double eyelid surgery is also known as _____, it is one of the most common plastic and cosmetic surgery.

2. Match the Chinese phrases with right English expressions.

红肿　　　　　bruise
淤青　　　　　cosmetics
化妆品　　　　drug metabolism
除皱术　　　　red and swollen
药物代谢　　　rhytidectomy

3. Arrange the following steps in the correct order.
(1) ice compress eyelid　(2) red, swollen and bruised eyes　(3) light diet
(4) avoiding frequent muscle movements　(5) knowing postoperative precautions
The correct order is: _____.

4. Translate the following sentences into English.
(1) 手术部位可能出现轻微红肿、淤青、疼痛感。
(2) 术后 6 小时内不要碰触操作部位, 24 小时后可外涂化妆品。
(3) 避免出现大笑、大哭等肌肉频繁运动。
(4) 术后 3 天不饮酒、吸烟,尽量不进行桑拿、泡温泉、高温瑜伽。
(5) 术后须注意清淡饮食。

Critical Thinking and Group Peer-assessment

1. Change the customer types or demands, redesigning the conversation.
2. Each group tries to display the dialogue according to the text.
3. After one group finishes the dialogue, others make comments.

 ## Knowledge Links

After surgery, new skin tissues around your eyes will settle in and it would take at least 3 months to make you look natural. The speed of the recovery also depends on the thickness of skin and the type of surgery method. The healing process will be entirely completed and you will look flawlessly natural about 6 months to 1 year after surgery.

 Task 4　Postoperative Follow-up

 ## Lead-in

The beauty consultant carries out postoperative follow-up after Ms. Li's double eyelid surgery.

Thinking and Talking

1. How to know postoperative state of double eyelid surgery?
2. The beauty consultant consults customers' postoperative state of double eyelid surgery.

 ## New Words

intermittently /ˌɪntəˈmɪtəntlɪ/	ad. 间歇地
liposuction /ˈlɪpəʊˌsʌkʃən/	n. 抽脂术
saline /ˈseɪlaɪn/	n. 生理盐水

Phrases and Expressions

surgical area	术区
eye-opening-and-closing exercises	睁闭眼运动
take out stitches	拆线
prevent infection	防止感染
wound healing	切口愈合

Intensive Reading

Text

(The beauty consultant is carrying out postoperative follow-up for Ms. Li.)

A: Beauty consultant　B: Ms. Li

A: Ms. Li, how do you feel after the surgery?

B: I feel a little pain.

A: It is normal to feel pain after surgery, but it will eventually go away.

B: OK. Is there any way to relieve the pain?

A: After surgery, you can make your pillow higher when you are asleep, use an ice compress on your surgical area within 2～3 days intermittently (alternating every 15 minutes), a hot compress 3 days later, and you can do more eye-opening-and-closing exercises.

B: After how long can I take out the stitches?

A: 5～7 days after surgery.

B: How about diet?

A: Ms. Li, please avoid spicy food in order to reduce irritation and scarring for 3 months.

B: OK, thanks for your postoperative follow-up. What can I do to cooperate with you?

A: Thank you, Ms. Li. Please accept and allow photography before and after surgery.

B: No problem.

A: Thanks a lot. I wish you an quick recovery and hope that you achieve the desired effects. Please comment on our service and we look forward to your valuable suggestions.

1. Fill in the blanks with the words given below, changing the form if necessary.

   > subcutaneous tissue(皮下组织)   swelling   infection   ice compress
   > nursing journal   liposuction   retrobulbar hemorrhage(球后出血)

   (1) Keep the wound clean after microplastic surgery in order to prevent _____.
   (2) _____ is loose connective tissue and adipose tissue(脂肪组织), it connects skin and muscles, it is also called the superficial fascia(浅筋膜).
   (3) _____ also known as liposuction fat loss and body sculpture.
   (4) You can make your pillow higher when you are asleep after surgery, _____ your surgical area within 2～3 days intermittently.
   (5) The surgeon suggests you keep _____ after surgery.

2. Match the Chinese phrases with right English expressions.

   冰敷   saline
   拆线   blepharoptosis
   生理盐水   take out stitches
   止疼   ice compress
   眼睑下垂   relieve pain

3. Arrange the following steps in the correct order.
   (1) eyes redness and swelling   (2) keeping clean after surgery   (3) wound healing
   (4) taking out stitches after surgery   (5) ice compress eyes
   The correct order is: _____.

4. Translate the following sentences into English.
   (1) 双眼皮手术当日伤口会有些疼痛。
   (2) 双眼皮手术应防止并发症的发生。
   (3) 术后出现大量出血应及时到医院复诊。
   (4) 术后可以用冰袋冷敷。
   (5) 双眼皮手术第二天可以做睁眼运动。

Critical Thinking and Group Peer-assessment

1. Change the customer types or demands, redesigning the conversation.
2. Each group tries to display the dialogue according to the text.
3. After one group finishes the dialogue, others make comments.

Knowledge Links

Taking care of yourself after a microplastic surgery is extremely important. The temperature and humidity of the operating room are suitable, the color is comfortable, the lighting is in line with the requirements of the operation, all kinds of rescue facilities are complete, so that patients have a sense of trust and security, along with maintaining a quiet and peaceful state of mind.

Unit 2

Beauty Care and Skin Ca

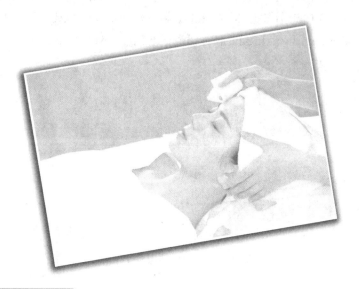

Learning Objectives

1. Learn the service process of skin care as well as preparations for professional facial care.
2. Be able to use the proper language in the process of facial care programs, and provide customers with advice on home treatment.

 Task 1 Introduction of Service Process

Ms. Wu has made an appointment for facial care at 10 a.m. After a beauty consultant makes a facial care program for her, a beautician leads her to a beauty room.

Thinking and Talking

1. What's the difference between the information provided by the beauty consultant and beautician to a customer who has facial care for first time and a regular customer?
2. After facial care, in what way could the beautician let the customer feel the positive effect of facial care?

New Words

slipper /ˈslɪpə/	n. 拖鞋
valuable /ˈvæljuəbl/	n. 贵重物品
necklace /ˈneklɪs/	n. 项链
earring /ˈɪərɪŋ/	n. 耳环
ingredient /ɪnˈɡriːdɪənt/	n. 配料
shiny /ˈʃaɪnɪ/	a. 有光泽的
elastic /ɪˈlæstɪk/	a. 灵活的；有弹性的
moisturize /ˈmɔɪstʃəraɪz/	v. 补水

Phrases and Expressions

make preparations	作准备
stay up	熬夜
cosmetic sprayer	美容蒸汽机

Intensive Reading

Text

(The beauty consultant has arranged for a beautician to give a facial for Ms. Wu. Now the beautician is guiding her to the beauty room.)

A: Beautician B: Ms. Wu

A: Hello, Ms. Wu. I am your beautician, Ada. I will assist you today.
B: Hello.
A: Please follow me this way. I have made preparations for you.
B: Thanks.
A: Please change your shoes here. The slippers are sterilized. Please feel free to use them.

B: OK.

(They enter the room)

A: Ms. Wu, please come in. You can put your belongings in the locker. Valuables such as your necklaces and earrings can also be kept in it.

B: OK, thanks.

A: This is a sterilized bathrobe. You can change your clothes and lie down. I will go prepare the ingredients for you and come back in five minutes.

B: Alright.

(After five minutes, the beautician waits outside and taps the door.)

A: Ms. Wu, may I come in?

B: Sure.

A: It's a little cold today. I will cover you with a quilt.

B: Thank you.

A: It's 10 o'clock now. I'll start the basic facial treatment for you. It will take about an hour.

B: OK, let's start.

A: Please relax. If you feel tired, you may take a nap.

B: OK, thank you.

(After an hour)

A: It's done. Would you like to lie down for a while or get up?

B: I am up! I have a dinner date with my friends.

A: Let me help you get up. Please change your clothes and get your belongings.

B: OK, thanks.

A: Look! Your skin looks shiny and elastic.

B: Really?

A: Yes. You had better moisturize your skin at home and go to bed early instead of staying up late at night. Remember to come here for facial care next week, if you are not busy.

B: OK.

(The beautician accompanies Ms. Wu to the reception desk.)

 **Practice**

Scanning Audio

1. Fill in the blanks with the words given below, changing the form if necessary.

| shiny   earring   relax   moisturize   elastic   ingredient   stay up |

(1) Please _____. If you feel tired, you can take a nap.
(2) I will go to prepare the _____ for you and come back in five minutes.
(3) Valuables such as necklaces and _____ can be kept in this locker.
(4) You had better _____ your skin at home.
(5) You had better go to bed early instead of _____ late at night.

2. Match the Chinese phrases with right English expressions.

美容蒸汽机　　　　beauty bed
红外线灯　　　　　infrared lamp
美容床　　　　　　black head remover
暗疮针　　　　　　cosmetic sprayer

3. Arrange the following steps in correct order.
(1) do facial care　(2) accompany the customer to the reception desk　(3) guide the customer to the beauty room　(4) test the skin　(5) make a plan of care program
The correct order is: _____.

4. Translate the following sentences into English.
(1) 请在前台签名,记得下周过来。
(2) 请稍等,我准备一下产品和设备。
(3) 请脱下您的外套,摘下您的耳环和项链。
(4) 您的皮肤有光泽和弹性。
(5) 您现在看上去神清气爽,神采奕奕。

Critical Thinking and Group Peer-assessment

1. Change the customer types or demands, redesigning the conversation.
2. Each group tries to display the dialogue according to the text.
3. After one group finishes the dialogue, others make comments.

Knowledge Links

Why Men Hate the Idea of Getting Facials?

There are many benefits of having a facial. However, there are many men who feel it is for the delicate skin and should be a service only enjoyed by women. Many men still believe it could make them look sensitive and feminine.

Why Men Like the Idea of Getting Facials?

There are other men who love the way they look and feel after a facial. Having a nice facial can also bring out confidence in a man, which many women might think is also very attractive and masculine.

Having a facial can remove years of buildup and oil that have been stored in pores. Having beautiful skin encourages a slight red flush on the cheeks, which usually helps to show off what sometimes we overlook, but is usually the most beautiful part of a man's face — their eyes.

Task 2 Preparation for Professional Facial Care

Lead-in

Ms. Zhang has made an appointment for facial care at 3 p.m. Before she begins facial care, a beauty consultant analyzes her skin and arranges a proper program for her.

Thinking and Talking

1. How many skin types are there? What are they?
2. What are the skin problems most likely to occur for different skin types?

New Words

toner /ˈtəʊnə/	n.	爽肤水,紧肤水
pat /pæt/	v.	轻拍
T-zone /tiːzəʊn/	n.	T区
mixed /mɪkst/	a.	混合的
replenish /rɪˈplenɪʃ/	v.	补充,装满
hydrating /ˈhaɪdreɪtɪŋ/	a.	保湿的;补水的
intensive /ɪnˈtensɪv/	a.	深层的

Phrases and Expressions

skin type 皮肤种类
stick with 坚持

Intensive Reading

Text

(This is the first time Ms. Zhang has come into the beauty salon. Now a beauty consultant is providing her with a skin analysis service.)

A: Beauty consultant   B: Ms. Zhang

A: Hello, I'm Mary, glad to assist you today. To better understand your skin condition, we will check your skin type and condition with this skin analysis apparatus. Do you know your skin type?

B: My skin might be dry. I pat on a lot of toner every day, but my skin is still dry.

A: Alright. Let's test and analyze your skin with the skin analysis apparatus.

(They test and analyze)

A: Look, you are dry on the cheeks, while your T-zone is oily, so you have mixed skin.

B: Oh, what should I do?

A: It's very important to replenish water and trap moisture at all times, especially for people with dry and mixed skin. It's better to use moisturizing masks daily or at least three times a week.

B: OK, I'll try my best.

A: According to your skin condition, I suggest our hydrating package with intensive moisturizing effects. It can repair your cells, increase elasticity and the glow of your skin.

B: How long until it take effect?

A: Just try once, and you will see an obvious moisturizing effect. It will remove the wrinkles on your face, if you stick with it.

B: That sounds great. I'll have that one then.

A: OK, just follow me.

Scanning Audio

1. Fill in the blanks with the words given below, changing the form if necessary.

   | mixed   obvious   wrinkle   skin type   cheek   repair   T-zone |

   (1) It will reduce the _____ on your face if you stick with it.
   (2) It can _____ your cells and increase elasticity and shine of skin.
   (3) You are dry on the _____, while your T-zone is oily, so you have mixed skin.
   (4) We can check your _____ and condition with this skin analysis apparatus.
   (5) Just use once, you can see moisturizing effects _____.

2. Match the Chinese phrases with right English expressions.

   中性皮肤   dry skin
   干性皮肤   oily skin
   敏感性皮肤   normal skin
   油性皮肤   mixed skin
   混合性皮肤   sensitive skin

3. Arrange the following steps in correct order.
   (1) explain the care program to the customer   (2) make a skin analysis for customer   (3) inquire the customer's demand   (4) recommend skin care programs to the customer   (5) confirm the facial care program
   The correct order is: _____.

4. Translate the following sentences into English.
   (1) 先做个彻底的面部清洁。
   (2) 我打过电话预约今天下午3:30做护理。
   (3) 最好每天或者至少每周使用三次保湿面膜。
   (4) 我们的活肤面膜可以促进血液循环,使皮肤更紧致。
   (5) 您可以选择手部按摩或者背部按摩。

Critical Thinking and Group Peer-assessment

1. Change the customer types or demands, redesigning the conversation.
2. Each group tries to display the dialogue according to the text.
3. After one group finishes the dialogue, others make comments.

Knowledge Links

How to Control Oily Skin?

10 dos or don'ts from dermatologists

(1) DO wash your face every morning, evening, and after exercise.

(2) DO choose skin care products that are labeled "oil free" and "noncomedogenic".

(3) DO use a gentle, foaming face wash.

(4) DON'T use oil-based or alcohol-based cleansers.

(5) DO apply moisturizer daily.

(6) DO wear sunscreen outdoors.

(7) DO choose oil-free, water-based makeup.

(8) DON'T sleep in your makeup.

(9) DO use blotting papers throughout the day.

(10) DON'T touch your face throughout the day.

 Standard Process of Facial Care

 Lead-in

After confirming the facial care program, a beautician leads Ms. Wu to a beauty room and begins the facial care service for her.

Thinking and Talking

1. What can the beautician talk to the customer when doing facial care?
2. Can people exfoliate the skin usually? Why?

 New Words

exfoliate /eksˈfəʊlɪeɪt/		v. 去角质
dirt /dɜːt/		n. 脏东西，污垢
nutrient /ˈnjuːtrɪənt/		n. 营养

lotion /ˈləʊʃən/	n. 乳液
faint /feɪnt/	a. 模糊
residue /ˈrezɪdjuː/	n. 残留物
essence /ˈesəns/	n. 精华
absorb /əbˈsɔːb/	v. 吸收

 Phrases and Expressions

makeup removal	卸妆
scrubbing cream	磨砂膏
essential oil	精油
facial mask	面膜
tonic lotion	爽肤水
day cream	日霜

 Intensive Reading

Text

(Ms. Wu lies down and the facial care begins.)

A: Ms. Wu   B: Beautician

A: What's this?

B: It's a cosmetic sprayer. It will help to open your facial pores, bring dirt out and put nutrients into the skin.

A: Oh, I see. What will you do during the facial?

B: First, I'll remove your makeup and select a gentle cleansing lotion to clean your face. After the cleansing step, I'll exfoliate your face with scrubbing cream. After that, I will massage your face with essential oil for 15～20 minutes, and then apply a facial mask. If you have time today, I can also provide eye care.

A: How long will that take?

B: Two hours altogether.

A: OK. I'll do both this time. I would like to sleep for a while. We don't need to talk.

B: I see. Have a good rest. During cleansing and massaging, please tell me if the pressure is too heavy or too light.

A: It's very comfortable, thank you.

(Two hours later, the beautician removes the mask from the customer.)

B: Can you see clearly?

A: No, I feel a bit faint.

B: Well. I will wipe up the residue in your eyes. How are you feeling now?

A: Much better.

B: Now, I have applied some tonic lotion, essence, day cream and sunscreen on your face. They will soon be absorbed. Would you like to get up or continue to lie down and rest?

A: Um, I would like to get up, and please give me my clothes.

B: Here you are.

A: Thank you. I had a really good rest today.

B: Wonderful! Please have a glass of water. You look very refreshed now.

A: Thank you for your excellent service.

B: Don't mention it. Please sign your name at the reception desk. Your beauty consultant is waiting for you. Remember to come here next week.

A: I will. See you next time.

B: See you.

Practice

Scanning Audio

1. Fill in the blanks with the words given below, changing the form if necessary.

> nutrient   pores   massage   faint   residue
> scrubbing cream   makeup   essential oil

(1) I'll remove your _____ and select a gentle cleansing lotion to clean your face.

(2) It's a cosmetic sprayer. It will help to open your facial _____.

(3) I will _____ your face with essential oil for 15~20 minutes.

(4) After the cleansing stage, I'll have an exfoliation with _____.

(5) I will wipe up the _____ in your eyes.

2. Match the Chinese phrases with right English expressions.

   黑眼圈   puffy eyes

   眼袋   panda eyes

   熊猫眼   black eye circle

   眼睛浮肿   dry eyes

   眼睛干涩   pouches

3. Arrange the following steps in correct order.

   (1) facial scrubbing   (2) facial cleansing   (3) facial makeup removing   (4) facial massaging   (5) applying facial mask   (6) essence importing   (7) facial toning

   The correct order is:_____.

4. Translate the following sentences into English.

   (1) 现在您脸上其实没有多少油脂，所以这次没必要为您去角质。

   (2) 如果您有时间的话，我还可以给您做眼部护理。

   (3) 您的鼻子上有一些黑头。

   (4) 我现在在您脸上涂一些爽肤水、精华液、日霜和防晒霜。

   (5) 您还记得上次去角质是什么时候吗？

Critical Thinking and Group Peer-assessment

1. Change the customer types or demands, redesigning the conversation.
2. Each group tries to display the dialogue according to the text.
3. After one group finishes the dialogue, others make comments.

Knowledge Links

Beauty Sleep Is Real

It is OK to cancel plans to say "I need my beauty rest" because beauty sleep is real. The quality and length of sleep you receive every night can have a profound impact on your skin's overall health. When we sleep our bodies recharge. Not only do our bodies recharge, our skin does as well. During sleep we heal, restore and eliminate toxins (毒素) from the skin. If sleep is compromised so is the body's ability to carry out these essential skin functions.

Part II Communication Skills in Cosmetology Domain   2-27

Home Treatment Advice

 Lead-in

Ms. Zhang chats with a beauty consultant about skin care at home.

Thinking and Talking

1. What are the differences between skin care at home and at beauty salon?
2. Is it important to do skin care at home? Why?

 New Words

complexion /kəmˈplekʃən/	n. 面色
regular /ˈreɡjʊlə/	a. 有规律的
soak /səʊk/	v. 浸泡

 Phrases and Expressions

foaming cleanser	洗面奶
night cream	晚霜
eye cream	眼霜
eye essence	眼部精华
sunscreen product	防晒产品

 Intensive Reading

Text

(The beauty consultant is having a daily chat with Ms. Zhang about her daily skin care routine.)

A: Beauty consultant   B: Ms. Zhang

A: Your complexion looks nice!
B: Really? That's good news for me.
A: How do you care for your skin at home? What products do you have?

B: After washing my face with warm water and foaming cleanser, I apply tonic lotion, then day cream or night cream.

A: Good. If you have heavy makeup, you'd better choose proper makeup removal lotion to wash your face until it's completely clean.

B: If I'm wearing makeup, I usually clean it thoroughly as soon as I come back home.

A: You should use makeup remover to clean your face first, even if you don't have makeup on, and then clean with foaming cleanser. The next step is to pat some tonic lotion on your face and neck to replenish water and moisture.

B: Do you think I should use any essence?

A: Of course, essence and eye cream or eye essence are necessary for people over thirty. They're applied before using day cream or night cream.

B: In summer, I use sunscreen products.

A: Well, it's a good habit to protect our skin from sunshine every day, not just in summer. In fact, regular exercise and a healthy lifestyle are more effective and practical. No expensive products can compare with them.

B: I see. Thank you.

A: You're welcome.

Practice

Scanning Audio

1. Fill in the blanks with the words given below, changing the form if necessary.

   | lifestyle   pat   complexion   night cream   essence   sunscreen products |

   (1) _____ some tonic lotion on face and neck to replenish water and moisture.

   (2) In summer, I use _____ to protect my skin from sunshine.

   (3) Applying_____ before using day cream or night cream are necessary for people over thirty.

   (4) Regular exercise and a healthy _____ are more effective and practical.

   (5) Your _____ looks nice!

2. Match the Chinese phrases with right English expressions.

   | 浓妆　　　　　light makeup
   | 淡妆　　　　　heavy makeup
   | 眼妆　　　　　cotton pads
   | 化妆棉　　　　eye makeup

3. Arrange the following steps in correct order.

   (1) face cream/eye cream (2) tonic lotion (3) face cleanser (4) facial essence/eye essence (5) makeup removal

   The correct order is: _____.

4. Translate the following sentences into English.

   (1) 每天涂防晒产品是个好习惯。
   (2) 经常化浓妆对皮肤不好。
   (3) 如果您化了浓妆,最好选择合适的卸妆乳洗脸直至完全清洁。
   (4) 就算不化妆,也应该先用卸妆乳洗脸,然后用泡沫洁面乳清洗。
   (5) 人们常常忽视颈部的皮肤护理。

Critical Thinking and Group Peer-assessment

1. Change the customer types or demands, redesigning the conversation.
2. Each group tries to display the dialogue according to the text.
3. After one group finishes the dialogue, others make comments.

Knowledge Links

The Mistakes Most Women Make

The biggest mistake we might make when removing makeup is that we are too stingy with our makeup removal products. We should soak cotton pads with eye makeup remover and apply a generous amount of cleanser on our faces.

The goal is to loosen up all that product we've smothered on our faces, and to do that properly, we need to be generous.

Another mistake we might make is we rub our eye area to remove makeup. This is a huge no-no. We should always pat, never rub, around the eye area.

And finally, we tend to forget to wash into our hairline and under our jawline, which is very important when we wear foundation.

Unit 3

Beauty Care and Body Care

Learning Objectives

1. Learn how to get the health information of a customer and introduce programs.
2. Be able to introduce service process and give aftercare tips in beauty salons.

 Task 1 Health Information of Customers

 Lead-in

A customer has recently suffered from sleeplessness and headaches, and her face starts to grow spots. She goes to the beauty salon, hoping this situation can be improved. A beauty consultant asks the customer about her health information.

Thinking and Talking

1. Do you know what items are written in the cover of a medical record?

2. What health information does a beauty consultant need to know from a customer?

New Words

form /fɔːm/	n. 表格
sleepless /ˈsliːplɪs/	a. 失眠的
gynecological /ˌgaɪnɪkəˈlɒdʒɪkəl/	a. 妇科的
menstrual /ˈmenstruəl/	a. 经期的
dysmenorrhea /ˌdɪsmenəˈrɪə/	n. 痛经
constipation /ˌkɒnstɪˈpeɪʃən/	n. 便秘

Phrases and Expressions

chest tightness	胸闷
fill out	填写
due to	由于
give birth	生孩子

Intensive Reading

Text

Health Form

Name: Zhang Lan		Gender: Female	Phone: 138 ********
Height: 165 cm		Weight: 65 kg	Age: 35
Marital status: Married		Child: 3 years old	
Major surgery: No			
Major illness: No			
Lifestyle	Eating habits: Eat on time; like to eat spicy food		
	Sports preferences: No frequent exercise		
	Sleep quality: Sometimes sleepless		
	Mental stress: Heavy		

续 表

Gynecological health	Menstrual period: 6 days; Cycle: 30 days; ● Dysmenorrhea
	○ Headache ● Low back pain ○ Chest tightness ● Sleepless ○ Constipation ○ Cold hands and feet
Facial Skin	Skin type: ● Oily ○ Normal ○ Dry ○ Combination ○ Sensitive
	Skin color: ○ Rosy ○ Fair ○ Pale yellow ● Dull
	Skin spots: ● Yes ○ No
	Skin relaxation: ● Yes ○ No

(Ms. Zhang is worried about the spots on her face. She decides to go to a beauty salon and ask the beauty consultant for help.)

A: Beauty consultant   B: Ms. Zhang

A: Hello, Ms. Zhang. I would like to ask you a few questions based on the form you have filled out.

B: OK.

A: Are you satisfied with your physical and skin condition?

B: I am not satisfied. I feel weak and my face has spots.

A: This may be due to giving birth.

B: Can it be improved?

A: Yes. You can try internal conditioning and some body care programs.

B: OK, I will try conditioning.

Scanning Audio

1. Fill in the blanks with the words given below, changing the form if necessary.

form spot surgery spicy sleepless stress menstruation conditioning

(1) _____ is a major stage of girls. It's one of the many physical signs that a girl is turning into a woman.

(2) When you have a toothache, do not eat _____ food.

(3) The first stage of the _____ was to join up the bones.

(4) Massage can help you relieve _____.

(5) Stress can lead to health problems, such as headaches and _____.

Part II Communication Skills in Cosmetology Domain

2. Match the Chinese phrases with right English expressions.

   失眠　　　　　　menstruation
   胸闷　　　　　　constipation
   月经　　　　　　dysmenorrhea
   便秘　　　　　　chest tightness
   痛经　　　　　　sleepless

3. Arrange the following steps in correct order.
   (1) greetings　(2) provide a program　(3) get back the health form　(4) fill out a health form
   The correct order is: _____.

4. Translate the following sentences into English.
   (1) 根据您填写的表格，我想问几个问题。
   (2) 您以前做过重大手术吗？
   (3) 您工作压力大吗？
   (4) 您晚上会失眠吗？
   (5) 这可能是由于您喜欢吃辛辣食物造成的。

Critical Thinking and Group Peer-assessment

1. Change the customer types or demands, redesigning the conversation.
2. Each group tries to display the dialogue according to the text.
3. After one group finishes the dialogue, others make comments.

Knowledge Links

Health Privacy

The privacy and security of patient health information is a top priority for patients and their families, health care providers and professionals, and the government. Health care providers and other key persons and organizations that handle your health information must protect it with passwords, encryption (加密技术), and other technical safeguards. These are designed to make sure that only the right people have access to your information.

Task 2 Programs Introduction

After childbirth, a customer has spots on her face. Besides, her belly has a bad shape, and her weight increases. She hopes the beauty salon can help her. A beauty consultant introduces to the customer some beauty and body care programs.

Thinking and Talking

1. Please list some beauty programs you know.
2. What will you say when introducing beauty programs?

 New Words

lighten /ˈlaɪtən/	v. 淡化
dredge /dredʒ/	v. 疏通
channel /ˈtʃænəl/	n. 经络
belly /ˈbelɪ/	n. 腹部
breast /brest/	n. 胸部
shaping /ˈʃeɪpɪŋ/	n. 塑形
gastrointestinal /ˌɡæstrəʊɪnˈtestɪnəl/	a. 肠胃的
discomfort /dɪsˈkʌmfət/	n. 不适
enhance /ɪnˈhɑːns/	v. 增强
meridian /məˈrɪdɪən/	n. 经络
slimming /ˈslɪmɪŋ/	n. 瘦身、减肥
excess /ɪkˈses/	a. 多余的
obese /əʊˈbiːs/	a. 肥胖的
pregnancy /ˈpreɡnənsɪ/	n. 怀孕;妊娠

Part II Communication Skills in Cosmetology Domain

 Phrases and Expressions

stretch mark   妊娠纹
take exercise   做运动

Text

Beauty Programs

No.	Program	Effect
1	Spot lightening	Repair spotted skin
2	Head treatment	Refresh your mind; improve sleep quality
3	Back treatment	Dredge internal channels; relieve stress
4	Belly shaping	Make the belly smaller
5	Breast care	Maintain the shape of breasts
6	Leg shaping	Reduce fat; shape legs
7	Hip shaping	Lift the hips
8	Gastrointestinal conditioning	Relieve stomach discomfort; enhance intestinal function
9	Stretch mark repairing	Repair stretch marks after childbirth
10	Meridian slimming	Remove excess body fat by meridians conditioning

(Ms. Zhang is obese after childbirth and hopes to improve her figure in a beauty salon.)

A: Beauty consultant   B: Ms. Zhang

A: Hello, Ms. Zhang. Can I help you?

B: Yes. Look at my belly. It is big after I gave birth to my child.

A: We have a belly shaping program. It helps to shape your belly.

B: Is the process comfortable?

A: Not really. You need exercise during the process. It is a bit tiring.

B: Is there any program which is more relaxing?

A: Yes, we have a meridian slimming program. It removes the excess fat by meridians conditioning. It is more comfortable.

B: What's the price?

A: 5,999 RMB.

B: How long does the whole process take?

A: Three months.

B: Sounds good. I think meridian slimming fits me more. Thank you.

A: You are welcome.

Scanning Audio

1. Fill in the blanks with the words given below, changing the form if necessary.

   | hip   shape   repair   program   excess   stretch   effect   release |

   (1) Tourism becomes an important way for people to _____ stress during their leisure time.

   (2) The main _____ of this product is to preserve health rather than treat illness.

   (3) The cream can moisturize and _____ your facial skin.

   (4) Proper dieting and exercising can help you to remove ugly _____ fat.

   (5) Stretch marks are caused by the skin being _____ during pregnancy.

2. Match the Chinese phrases with right English expressions.

   淡斑　　　　　slimming
   臀部塑形　　　spot lightening
   睡眠质量　　　spotted skin
   有斑皮肤　　　hip shaping
   瘦身　　　　　sleep quality

3. Arrange the following steps in correct order

   (1) assign a beautician　(2) create a file　(3) introduce a beauty program　(4) understand the customer's needs

   The correct order is: _____.

4. Translate the following sentences into English.

   (1) 这个项目可以祛除产后妊娠纹。
   (2) 这个项目效果不错。
   (3) 请问整个过程需要多长时间？
   (4) 我觉得经络减肥更适合我。
   (5) 头部疗法可以释放压力。

Critical Thinking and Group Peer-assessment

1. Change the customer types or demands, redesigning the conversation.
2. Each group tries to display the dialogue according to the text.
3. After one group finishes the dialogue, others make comments.

 Knowledge Links

Weight Loss Calories

If you want to lose weight, keep in mind that a deficit of 500 calories per day will help you to lose about 1 pound (454g) per week. A calorie deficit of 1,000 calories per day can result in a 2-pound weight loss per week. Cutting more than 1,000 calories from your diet may do more harm than good. Very low-calorie diets should only be followed under a doctor's supervision.

Task 3 Health Care Service Process

 Lead-in

Tired of work and heavy chores, a customer has often felt back pain recently. She has reserved a massage program at a beauty salon. The beautician lets the customer understand the massage service process.

Thinking and Talking

1. What materials are needed for massage?
2. What is the process of massage?

 New Words

detox /ˈdiːtɒks/   a. 排毒的

flow /fləʊ/	n. 流动
lymphatic /lɪmˈfætɪk/	a. 淋巴的
grease /ɡriːs/	n. 油脂
circular /ˈsɜːkjʊlə/	a. 圆圈的
spine /spaɪn/	n. 脊柱
technique /tekˈniːk/	n. 手法
acupoint /ˈækjʊpɔɪnt/	n. 穴位

Phrases and Expressions

lymphatic system	淋巴系统
excess water	多余水分
circular massage	打圈按摩

Intensive Reading

Text

(Lucy has reserved a back massage. She goes to the beauty salon.)

A: Beautician   B: Lucy

A: Hello Lucy. Your appointment today is a back detox massage.

B: Yes. This is my first time doing this massage. Would you please explain it first?

A: The purpose of this massage is to increase smooth flow of the lymphatic system and eliminate excess water and toxins in the body. It takes about one hour.

B: OK.

A: I am now cleaning your back to remove dirt and grease from your skin.

B: I have been sweating a lot recently. The skin must be oily.

A: I am now exfoliating the skin on your back for further cleansing.

B: OK, it feels comfortable when you are doing a circular massage.

A: It is done. Now I am pressing your skin to help you relax more fully.

B: OK. Thanks.

A: The next step is a whole back massage for about 30 minutes. I am now applying a lymphatic detox technique to massage you, removing toxins from your body.

B: It aches, but in a comfortable way.

A: Now I am massaging your waist at the back. The force is slightly stronger when I am pressing on your spine. Now, how do you feel? Would you like stronger pressure?

B: I think it's alright.

A: I'm now massaging an important acupoint in your waist. Is it comfortable?

B: It's a little painful. Softer, please.

A: OK.

(About half an hour later)

A: The massage is over. Now I will clean the skin on your back for the last time. How do you feel now?

B: I feel much more relaxed.

A: That's good. Please put on your clothes. Keep warm.

B: OK. Thank you.

A: You are welcome. Please come here to take your bag.

Scanning Audio

1. Fill in the blanks with the words given below, changing the form if necessary.

| massage   technique   lymphatic   toxin   spine   sweat   exfoliation   eliminate |

(1) If you often sit in the wrong position, your _____ might be damaged.

(2) The disease releases _____ into the bloodstream.

(3) _____ is used to relax muscles, relieve stress and improve the circulation.

(4) The _____ system plays an important role in the immune functions of the body.

(5) _____ takes the dead cells out of your skin and gives it back the healthy look.

2. Match the Chinese phrases with right English expressions.

排除毒素   lymphatic system
淋巴系统   cleansing
清洁   eliminate toxins
出汗   grease
油脂   sweat

3. Arrange the following steps in correct order.

(1) exfoliating   (2) grease cleaning   (3) massaging   (4) pressing

The correct order is: _____.

4. Translate the following sentences into English.

(1) 您今天预约的是背部按摩。

(2) 这个按摩的作用是促进淋巴系统的畅通。
(3) 我现在开始给您的背部进行清洁。
(4) 您感觉怎么样？需要加大力度吗？
(5) 我感觉有点疼，请轻一点。

Critical Thinking and Group Peer-assessment

1. Change the customer types or demands, redesigning the conversation.
2. Each group tries to display the dialogue according to the text.
3. After one group finishes the dialogue, others make comments.

Knowledge Links

What Is Hot Stone Massage?

A hot stone massage is a massage that uses smooth, flat, and heated rocks placed at key points on the body. Stones are placed into an electric slow-cooker or a purpose-built device which is filled with water. The water is typically heated to 50.0~52.8℃. Once the stones have heated sufficiently, some are placed onto specific points on the body (such as the back, hands, etc.) and others are held by the massage therapist and used to work the muscles.

Task 4 Health Care Tips

Lead-in

Some students go to the canteen to eat right after they finish running 800 meters. Some patients go to play basketball with friends right after they discharge from the hospital after surgery. Do you think their behaviors are correct? Why or why not?

Thinking and Talking

1. Have you ever had moxibustion? What is the function of moxibustion?

2. What should you pay attention to after moxibustion?

New Words

moxibustion /ˌmɒksɪˈbʌstʃən/	n. 艾灸
scraping /ˈskreɪpɪŋ/	n. 刮痧
cupping /ˈkʌpɪŋ/	n. 拔罐
scratch /skrætʃ/	v. 抓挠
blister /ˈblɪstə/	n. 水泡
alcoholic /ˌælkəˈhɒlɪk/	a. 酒精的
itchy /ˈɪtʃɪ/	a. 发痒的

Phrases and Expressions

take a shower	洗澡
take a break	休息
come up	出现

Intensive Reading

Text
Health Care Tips

I. Tips after moxibustion
1. Quietly rest for 15 minutes.
2. Do not take a shower within 4 hours.
3. Keep warm.
4. Do not eat spicy food.
5. Do not take strenuous exercise.

II. Tips after scraping
1. Take a break for more than an hour.
2. Take a shower 12 hours later.
3. Drink warm water.
4. Do not use a fan, so as not to catch a cold.

5. Rest for 5~7 days before another scraping.

III. Tips after cupping

1. Rest for 15~20 minutes.

2. Do not take a cool shower within 30 minutes.

3. Drink a cup of warm water after cupping.

4. To avoid infection, do not scratch.

5. Go to hospital in time when blisters come up.

IV. Tips on SPA

1. Properly cool down before the SPA.

2. Drink plenty of water before and after the SPA.

3. Go to the SPA half an hour after a meal.

4. Do not have any alcoholic drinks 3 hours before the SPA.

5. Rest for half an hour before leaving.

6. People with heart disease, skin disease, high blood pressure or pregnancy should not have a SPA.

1. Fill in the blanks with the words given below, changing the form if necessary.

| tip   strenuous   blister   alcoholic   SPA   scratch   acupuncture   infection |

　(1) _____ salons and fitness centers are the ideal places to relax.

　(2) _____ treatment is gentle, painless and relaxing.

　(3) Do not _____ the wound when it is itchy.

　(4) Do not take _____ exercise immediately after a meal.

　(5) _____ drinks act as a poison to a baby.

2. Match the Chinese phrases with right English expressions.

　酒精饮品　　　heart disease
　刮痧　　　　　alcoholic drinks
　艾灸　　　　　cupping
　拔罐　　　　　scraping
　心脏病　　　　moxibustion

3. Arrange the following steps in correct order.

　(1) clean the mess　(2) prepare tools　(3) do cupping　(4) offer tips

　The correct order is: _____.

4. Translate the following sentences into English.
   (1) 12小时后再洗澡。
   (2) 不要吹风扇,以免着凉。
   (3) 不要抓挠,以免感染。
   (4) 拔罐后饮一杯温开水。
   (5) 请勿饮用任何含酒精的饮品。

Critical Thinking and Group Peer-assessment

1. Change the customer types or demands, redesigning the conversation.
2. Each group tries to display the dialogue according to the text.
3. After one group finishes the dialogue, others make comments.

Knowledge Links

What Are the Benefits of Acupuncture?

A growing body of clinical research is discovering how the body responds to acupuncture and its benefits for a wide range of common health conditions. A lot of people have acupuncture to relieve specific aches and pains, such as osteoarthritis(骨关节炎) of the knee, headaches and low back pain, or for common health problems like an overactive bladder(膀胱过动症). Other people choose acupuncture when they can feel their bodily functions are out of balance, but they have no obvious diagnosis. And many have regular treatments because they find it so beneficial and relaxing.

Discussion Questions

1. What attitudes should we have when serving the customers?
2. What do you think customers care about when choosing a beauty program?

Part III Introduction of Cosmetics

Unit 1
Facial Care Products

> **Learning Objectives**
>
> 1. Learn to recommend facial care products to customers in English according to their skin characteristics and needs.
> 2. Be able to understand and translate the labels of facial care products.

Task 1 Facial Cleanser

Lead-in

Facial cleanser is one of the most important products in your skin care process. It works to remove dirt, makeup and other impurities to leave your skin fresh and clean. Cleansing is the first step in skin care. However, there are some people who don't know the importance of it, and Ms. Li is one of them. She thinks that it is unnecessary to wash her face with facial cleanser as long as she doesn't have makeup.

Thinking and Talking

1. What do you think of Ms. Li's opinion?
2. Would you use facial cleanser to wash your face every day?

New Words

efficacy /ˈefɪkəsɪ/	n.	功效,效力
direction /dɪˈrekʃən/	n.	用法说明
squeeze /skwiːz/	v.	挤压;压榨
impurity /ɪmˈpjʊərɪtɪ/	n.	杂质;不纯;不洁
sweep /swiːp/	v.	扫除
nutrition /njʊˈtrɪʃən/	n.	营养
irritate /ˈɪrɪteɪt/	v.	刺激
pH	n.	酸碱值

Phrases and Expressions

applicable skin	适用皮肤
rinse thoroughly	彻底清洗
amino acid	氨基酸

Intensive Reading

Text

Product name: Facial Cleanser.

Applicable skin: For all types of skin, including sensitive skin.

Efficacy: Deeply cleaning and moisturizing skin.

Directions:

Squeeze the cleanser into your hands, massage your face, and rinse thoroughly.

Conversation

(Ms. Li is 28 years old. It is the first time she will have facial care in the beauty salon. During the process, she proactively asks the beautician about the necessity of using facial cleanser.)

Part III Introduction of Cosmetics

A: Beautician B: Ms. Li

A: Ms. Li, first of all, I will clean your face.
B: Do I need to use facial cleanser every time I wash my face?
A: Yes, Ms. Li. Cleansing is the first step in skin care. It removes dirt, makeup and other impurities from your face as if it were "sweeping the floor".
B: I thought I wouldn't need a facial cleanser if I didn't have makeup on.
A: No, only after cleaning the skin thoroughly every time can the skin absorb the nutrition in the ingredients better from the skin care. So it is very important to use facial cleanser.
B: Which type of facial cleanser should I use?
A: Traditional, soap-based facial cleanser irritates skin, which may cause skin sensitivity. It is recommended that you use an amino acid facial cleanser, which is soap-free, and has a pH that is close to that of the skin. While cleansing, it does not irritate the skin and is suitable for all types, including sensitive skin.
B: OK, I'll buy one and try it out.
A: All right.

 *Practice*

Scanning Audio

1. Fill in the blanks with the words given below, changing the form if necessary.

irritate efficacy massage applicable cleanser nutrition soap amino acid

(1) You should use _____ before moisturizing your face.
(2) Too much massage can _____ the skin and cause redness.
(3) Washing your face helps your skin absorb the _____ from the following skin care.
(4) The _____ skin type of this cleanser is oily skin.
(5) The _____ of facial cleanser is to deeply clean the grease and dirt on the skin.

2. Match the Chinese phrases with right English expressions.

面霜   facial cleanser
化妆水   eye cream
洗面奶   essence
精华   facial cream
眼霜   lotion
乳液   toner

3. Arrange the following steps in correct order.
   (1) essence (2) facial cleanser (3) eye cream (4) toner (5) facial cream
   The correct order of basic facial care is: _____.
4. Translate the following sentences into English.
   (1) 用洗面奶洗脸是皮肤护理的第一步。
   (2) 敏感肌肤适合用氨基酸洗面奶洗脸。
   (3) 每天洁面两次,早晚各一次。
   (4) 本产品未添加任何皂基。
   (5) 我们为您推荐一位经验丰富的美容师为您服务。

Critical Thinking and Group Peer-assessment

1. Change the customer types or demands, redesigning the conversation.
2. Each group tries to display the dialogue according to the text.
3. After one group finishes the dialogue, others make comments.

Knowledge Links

How to Choose a Facial Cleanser

The first step to choose the right cleanser is to know your skin type. Do you have oily skin, dry skin, combination skin, sensitive skin, or acne prone skin? This information is critical because if you use a cleanser for oily skin when you actually have dry skin, you'll strip your skin of the moisture it needs. If you use an acne cleanser on sensitive skin, you risk irritation.

https://www.bioelements.com/blogs/blog/choose-facial-cleanser

Task 2 Exfoliating Gel

Lead-in

Regular exfoliation helps to keep your skin looking fresh, radiant and smooth and stimulates new skin cells to grow. Ms. Li has never used any exfoliators, therefore her face is dark and has black heads.

Thinking and Talking

1. Is exfoliation necessary for all types of skin?
2. Do we need to exfoliate every day?

New Words

radiant /ˈreɪdɪənt/	a. 容光焕发的；光芒四射的
smooth /smuːð/	v. 使光滑
stimulate /ˈstɪmjʊleɪt/	v. 刺激；鼓舞，激励
cutin /ˈkjuːtɪn/	n. 角质；表皮素
appropriate /əˈprəʊprɪət/	a. 适当的；合适的
neglect /nɪˈɡlekt/	v. 疏忽，忽视；忽略

Phrases and Expressions

black head	黑头
fruit acid	果酸
dead skin cells	死皮细胞

Intensive Reading

Text

Product name: Exfoliating gel.

Efficacy: Removing cutin gently, clearing pores and replenishing water to skin.

Suggestions of usage cycle:

Oily skin: once or twice a week.

Dry skin: once every 10 days.

Normal skin: once every 15 days.

Mixed skin:

T-zone: once or twice a week.

U-zone: once every 10 days.

Directions: Apply appropriate amount to the part to be exfoliated, massage in circles for 1~2 minutes, then rinse thoroughly.

Conversation

(The beautician has cleaned Ms. Li's face and is ready to do exfoliating care for her.)

A: Beautician   B: Ms. Li

A: Ms. Li, can I exfoliate your skin today?

B: Well, I often neglect this part in my skin care routine.

A: I'll use this exfoliating gel. Its natural ingredients of fruit acid can gently remove dead skin cells from your face without irritation. When you come to us for skin care every month, we will use the instrument to do a thorough cleaning for you. You can also do it at home by yourself.

B: That's good. How do you use it?

A: Apply an appropriate amount to your face, massage in circles for 1~2 minutes, then rinse thoroughly. It is very important and necessary to do basic facial care after exfoliating. Since you have oily skin, you'd better do the exfoliation once or twice a week.

B: OK.

A: Remember that exfoliating too often can also damage your skin.

B: Oh, I see.

Practice

Scanning Audio

1. Fill in the blanks with the words given below, changing the form if necessary.

exfoliate   neglect   stimulate   appropriate   satisfied   cutin   instrument   gently

(1) Dry skin people should _____ their skin once every 10 days.

(2) Take _____ amount of gel and massage in circles for 2 minutes, then wash off.

(3) This gel can _____ exfoliate your skin.

(4) Li is _____ with her skin: bright, smooth, and hydrated.

(5) We will use professional _____ to serve your skin care.

Part III Introduction of Cosmetics

2. Match the Chinese phrases with right English expressions.

    按摩   exfoliate
    去角质   massage
    吸收   gel
    啫喱   absorption

3. Arrange the following steps in correct order.

    (1) massaging   (2) rinsing thoroughly   (3) basic facial care   (4) applying

    The correct order of exfoliation is：_____.

4. Translate the following sentences into English.

    (1) 油性皮肤每周应该进行1~2次去角质护理。
    (2) 过多的角质会导致皮肤暗沉。
    (3) 千万不要忽略去角质这个护肤环节。
    (4) 去角质护理之后一定要进行面部基础护理。
    (5) 定期去角质是很有必要的。

Critical Thinking and Group Peer-assessment

1. Change the customer types or demands, redesigning the conversation.
2. Each group tries to display the dialogue according to the text.
3. After one group finishes the dialogue, others make comments.

Knowledge Links

Skin Over-exfoliation Effects

- Very dry patches of skin
- Rough, bumpy skin
- Uneven skin tone
- Random breakouts on various parts of my face
- Skin that felt like sandpaper
- Tough, hardened skin

https://www.mywomenstuff.com/2016/06/over-exfoliate-skin-repair/

Task 3 Toner

We should use toner to hydrate the skin after the skin has been cleansed. If your skin tends to be on the dry side, toners may add an extra level of hydration. Ms. Li always feels dry in the face in winter.

Thinking and Talking

1. Toner has many other benefits. Do you know what they are?
2. How is toner used?

New Words

hydrate /ˈhaɪdreɪt/	v. 使成水化合物；补水
specification /ˌspesɪfɪˈkeɪʃn/	n. 规格
balance /ˈbæləns/	v. 使平衡
can /kæn/	n. 容器
spray /spreɪ/	v. 喷；喷射

Intensive Reading

Text

Product name: Toner.

Specification: 500 mL.

Applicable skin: For all types of skin, including sensitive skin.

Efficacy: Hydrating and moisturizing, cleansing skin for a second time and helping to balance water and oil.

Conversation

(Exfoliation has been done for Ms. Li, and she continues to enjoy the skin care program.)

A: Beautician   B: Ms. Li

A: Ms. Li, let's begin to apply toner on you.

B: All right.

A: Even though we have rinsed the skin with water, there are still excess oils and dead skin cells left behind, as well as impurities in the water that can also stay on our skin. The toner can help us to cleanse the skin for a second time.

B: I see.

A: Moreover, toner can help open skin pores and absorb nutrients from other products more quickly.

B: OK.

A: That's why we will apply toner to your skin at the end of each step. Hydrating and moisturizing.

B: Oh! Please give my skin more water.

A: You can put the toner into a small watering can and spray it when you are at work during the day, hydrating and helping to balance skin's water and oil. You can also use it before having a facial mask in the evening in order to open your skin pores.

B: OK. There are so many benefits. Please let me have a look later on.

Practice

Scanning Audio

1. Fill in the blanks with the words given below, changing the form if necessary.

> specification rinse hydrate impurity nutrient
> effective facial spray facial mask

(1) The efficacy of toner is _____ and moisturizing skin.
(2) What is the _____ of this product, 100 mL or 150 mL?
(3) Toner can help the skin absorb _____ from the following products.
(4) What is the _____ method to hydrate for skin?
(5) This _____ toner can moisturize your skin at any time.

2. Match the Chinese phrases with right English expressions.

规格　　　　primer
隔离乳　　　specification
面膜　　　　spray
喷雾　　　　facial mask

3. Arrange the following steps in correct order.

(1) makeup (2) sunscreen (3) primer (4) basic skin care

The correct order of daytime facial care is: _____.

4. Translate the following sentences into English.

(1) 水里有很多杂质。
(2) 这款爽肤水规格是 300 mL。
(3) 本品涂抹之后无须清洗。

（4）爽肤水的使用方法很多。

（5）你知道如何保持肌肤的水油平衡吗？

Critical Thinking and Group Peer-assessment

1. Change the customer types or demands, redesigning the conversation.
2. Each group tries to display the dialogue according to the text.
3. After one group finishes the dialogue, others make comments.

Knowledge Links

Primer

If we're being honest here, primer is one of those products we used to skip. But, when you're applying makeup yourself (and wearing it all day long), you will definitely notice the difference between using primer and going without it. Primers sit on the surface of the skin and create a barrier between skin and makeup. Primer helps keep skin looking smooth, pores looking smaller, and delivers a subtle golden glow. It's a must during the summer months especially, as it will keep you from sweating off your makeup.

https://laurenconrad.com/blog/2016/02/beauty-beginners-the-right-way-to-prime-your-face-for-makeup/

Task 4 Eye Cream

Lead-in

The skin around the eyes is sensitive and the cuticle is thin. If not properly maintained, it is often the first place where wrinkles appear. Ms. Wang has dry skin. There are fine lines on her face. Especially when she is laughing, the wrinkles around her eyes are more obvious.

Thinking and Talking

1. What are the benefits of eye cream?

Part III Introduction of Cosmetics   3-13

2. How should you apply eye cream?

New Words

cuticle /ˈkjuːtɪkl/	n. 角质层；表皮；护膜
intensity /ɪnˈtensɪtɪ/	n. 强度
dot /dɒt/	v. 打上点
antioxidant /ˌæntɪˈɒksɪdənt/	n. 抗氧化剂
extract /ˈekstrækt/	n. 提取物

Phrases and Expressions

dynamic fold	动态皱纹
true wrinkle	真性皱纹
crow's feet	鱼尾纹

Intensive Reading

Text

Product name: Eye cream.

Specification: 15 mL.

Efficacy: Improving wrinkles and reducing dynamic folds.

Directions: Apply appropriate amount around the eyes twice daily before using other skincare products.

Conversation

（Ms. Wang is 35 years old. It is her first time having facial care in the beauty salon. A beautician is ready to be at her service.）

A: Beauty adviser   B: Ms. Wang

A: Hello, Ms. Wang! Your skin test showed that your skin is dry. Therefore, hydration and moisturizing should be in your daily routine of skin care.

B: I'm using moisturizing cosmetics, but I still feel dry on my face. Especially when I am laughing, there are obvious fine lines around my eyes.

A: OK, I am doing our professional hydration care for you, which can help to keep your skin from water shortage. However, since it's winter now, the water shortage of skin will be more serious, especially the skin around the eyes. You need to take good care of that part.

B: I haven't used eye cream before.

A: The skin around the eyes is sensitive and the cuticle is thin. It is often the first place where wrinkles appear. Eye cream can improve fine lines around the eyes and slow down the process of it becoming true wrinkles.

B: How long will it take to see the effects?

A: Only when you are using eye cream for a long time can you see the effects. So you should stick to using it.

B: How do I use it?

A: Use it every morning and evening before the facial cream. The method of applying eye cream is different from that of facial cream, and the intensity of massage is also different. Dot the skin around the eyes with eye cream. Pat gently until it is absorbed. Do not pull, otherwise it will easily lead to crow's feet.

B: OK.

Scanning Audio

1. Fill in the blanks with the words given below, changing the form if necessary.

improve   reduce   obvious   method   professional   routine   strength   moisturize

(1) It can _____ your skin and keep your skin full of water.

(2) You need to use anti-aging and antioxidant products to _____ your skin.

(3) Our beauty salon will provide you with the most _____ services.

(4) Applying eye cream is part of your daily skin care _____ .

(5) The wrinkles around my eyes are _____ . How should I remove them?

2. Match the Chinese phrases with right English expressions.

动态细纹　　　　fine lines

皱纹　　　　　　ingredient

成分　　　　　　wrinkle

细纹　　　　　　dynamic fold

3. Arrange the following steps in correct order.

(1) applying eye shadow　(2) drawing eyeliner　(3) using mascara　(4) using eyelash curler

The correct order of eye makeup is: _____.

4. Translate the following sentences into English.
   (1) 如何改善我的肌肤呢？
   (2) 点涂在眼周肌肤上,轻拍直至被吸收。
   (3) 冬季皮肤缺水会更严重,因此要小心呵护眼周的肌肤。
   (4) 本款产品可以帮助抚平细纹。
   (5) 坚持使用,效果更好。

Critical Thinking and Group Peer-assessment

1. Change the customer types or demands, redesigning the conversation.
2. Each group tries to display the dialogue according to the text.
3. After one group finishes the dialogue, others make comments.

Knowledge Links

How to Apply Eye Shadow

Step 1: Choose the right brush to create the look you want.
- A large shadow brush is best for lightly coating your eyelids with a thin layer of natural looking color.
- A medium brush provides more coverage.
- A small brush provides the greatest color intensity and precision.
- Angle brushes are best for lining and detailing.

Step 2: Pick three eyeshadows that are similar in color.

Step 3: Sweep an eyeshadow brush through some shadow and tap off the extra.

Step 4: Brighten your inner eyelid and brow bone with the light shade.

Step 5: Create depth in your crease with the darkest shadow.

Step 6: Use the middle shade to connect the two others.

Step 7: Blend your eyeshadow together.

https://www.wikihow.life/Apply-Eye-Makeup

Task 5 Moisturizing Cream

Lead-in

There are many kinds of moisturizing creams with different functions, such as whitening, anti-oxidation, hydrating and tightening. The selling points of each product are different from each other. Customers can choose reasonably according to their own needs. You can see nasolabial folds on Ms. Wang's face. What kind of moisturizing cream should she use?

Thinking and Talking
1. What kinds of cosmetics have antioxidant efficacy?
2. Can moisturizing cream improve false wrinkles?

New Words

tighten /ˈtaɪtn/	v.	使变紧；绷紧
palm /pɑːm/	n.	手掌
laxity /ˈlæksətɪ/	n.	松弛；放纵
molecule /ˈmɒlɪkjuːl/	n.	分子；微粒
feature /ˈfiːtʃə/	n.	特色，特征
activate /ˈæktɪveɪt/	v.	使活动；刺激
premature /ˈpremətjʊə/	a.	早产的；比预期早的

Phrases and Expressions

moisturizing cream	润肤霜
eye contour	眼部轮廓
nasolabial fold	法令纹
skin aging	皮肤老化
fake wrinkle	假性皱纹

Part III Introduction of Cosmetics

Text

Product name: Moisturizing cream.

Efficacy: Highly antioxidant efficacy. Moisturizing and smoothing your skin. Improving wrinkles and tightening skin.

Specification: 60 mL.

Directions: Take appropriate amount and warm the cream for a few seconds between fingers or palms. Gently press into face, eye contour and neck.

Conversation

(The beautician continues to provide facial care for Ms. Wang.)

A: Beautician B: Ms. Wang

A: Ms. Wang, let's continue with your skin care.

B: All right. In addition to the fine lines around my eyes, I think there are also wrinkles on my face, especially the nasolabial folds.

A: The nasolabial folds are caused by skin laxity. When people get older, the skin becomes loose, which is a sign of skin aging.

B: What cosmetics should I use?

A: The moisturizing cream I am using for you contains small molecules. It has features like good absorption, moisturizing and locking in water, promoting improved metabolism and activating cells, etc. It contains highly antioxidant ingredients which can moisturize and smooth, improve wrinkles and tighten skin. Wrinkles are not likely to appear in highly moisturized skin.

A: That sounds good. How do I use it?

B: Use it every morning and evening, after the eye cream. Apply it to the face after warming it with your fingers or palms.

A: OK. Show me later, please.

Scanning Audio

1. Fill in the blanks with the words given below, changing the form if necessary.

| moisturize palm efficient antioxidant |
| molecule promote laxity injection |

(1) Warm the cream between your _____ until it turns to be transparent.
(2) This moisturizer has smaller _____, so it is easy to be absorbed.
(3) The active ingredients of this moisturizer can _____ metabolism.
(4) True wrinkles can be improved by using _____, etc.
(5) Moisturizing cream has the function of _____ and can remove false wrinkles.

2. Match the Chinese phrases with right English expressions.

抗氧化剂　　　　molecule
眼周　　　　　　antioxidant
新陈代谢　　　　eye contour
分子　　　　　　metabolism

3. Arrange the following steps in correct order.
(1) toner (2) eye cream (3) cleanser (4) moisturizing cream (5) essence
The correct order of night facial care is: _____.

4. Translate the following sentences into English.
(1) 本品具有抗氧化作用,有效防止皮肤衰老。
(2) 分子颗粒小的化妆品很容易被肌肤吸收。
(3) 法令纹是由于皮肤松弛引起的。
(4) 含水量高的肌肤就不容易产生细纹。
(5) 动态皱纹属于假性皱纹。

Critical Thinking and Group Peer-assessment

1. Change the customer types or demands, redesigning the conversation.
2. Each group tries to display the dialogue according to the text.
3. After one group finishes the dialogue, others make comments.

Knowledge Links

How to Erase Fine Lines & Wrinkles on Your Face

Step 1: Try AHAs. Try creams and lotions with alpha-hydroxy acids, which remove dead skin cells, reduce lines and age spots, and exfoliate. Start with a low dose every other day and work up to daily use to avoid irritation.

Step 2: Use retinol. Use topical retinol, a powerful antioxidant derived from vitamin A, proven to promote collagen production and plump skin while ironing out wrinkles.

Step 3: Look for antioxidants. Look for moisturizers and products with other important antioxidants, including vitamins A, C, and E, alpha-lipoic acid, acai oil, and coenzyme Q10.

Step 4: Consider hyaluronic acid. Consider products with hyaluronic acid, something your body produces naturally but slows with age. Hyaluronic-acid-based injectable fillers instantly plump skin and stimulate collagen production.

Step 5: Drink healthy. Swap coffee for green tea and white wine for red wine. Both red wine and green tea contain anti-aging antioxidants and polyphenols.

Step 6: Wear sunscreen. Wear sunscreen year-round on your face, neck, and chest. Sun is the number one cause of premature aging.

https://www.howcast.com/videos/326567-how-to-erase-fine-lines-wrinkles-on-your-face

Task 6 Sunscreen

Lead-in

Sun exposure is the number one cause of premature aging. No matter what season it is, you must wear sunscreen as long as you go out. Ms. Li is going to take her children to Thailand for a tour. She doesn't know what kinds of sunscreen products she should buy for seaside tour and for children. She's confused.

Thinking and Talking

1. What products would you recommend to her?
2. Do you know the harm of not wearing sunscreen?

New Words

principle /ˈprɪnsɪpl/	n. 原理
burden /ˈbɜːdn/	n. 负担；责任
SPF	abbr. 防晒系数（Sun Protect Factor）

PA	*abbr.* 防御 UVA 能力（Protection Grade of UVA）
defense /dɪˈfens/	*n.* 防卫，防护
pigmentation /ˌpɪgmənˈteɪʃən/	*n.* 色素淀积
waterproof /ˈwɔːtəpruːf/	*a.* 防水的，不透水的
sweatproof /ˈswetpruːf/	*a.* 防汗的，不透汗水的

Phrases and Expressions

physical sunscreen	物理防晒
chemical sunscreen	化学防晒
pregnant woman	孕妇

Intensive Reading

Text

According to the principle of sunscreen, it can be divided into two types, physical sunscreen and chemical sunscreen. In comparison, physical sunscreen causes fewer burdens on the skin, and is not easy to cause skin sensitivity. Therefore, it is more suitable for pregnant women and children.

Directions:

1. Apply sunscreen after your daily skin care routine but before applying makeup.
2. Whether it being cloudy or sunny, you must be protected from the sun as long as you go out.
3. Apply sunscreen 10 to 20 minutes before going out.
4. Be sure to thoroughly clean the sunscreen before going to bed.

Shopping Suggestions:

1. Choose different types of sunscreen according to different skin types. Water-based sunscreen product is suitable for oily skin; sunscreen cream is suitable for dry skin; and sunscreen lotion is suitable for all types of skin.
2. The higher the SPF, the longer the protection, the more "＋" after PA, the better the defense.

Part III Introduction of Cosmetics   3-21

Conversation

(Ms. Wang continues to enjoy her facial care.
She consults the beautician about sunscreen products.)

A: Beautician   B: Ms. Wang

A: Ms. Wang, I will apply sunscreen cream for you.
B: OK. I am going to take my kid to Thailand in a few days.
A: Then using sunscreen to protect your skin will be very important. Otherwise, it will be easy to cause pigmentation and form spots after getting sun burnt.
B: Is that serious?
A: Yes. Sun exposure is the number one cause of premature aging. In fact, we must wear sunscreen as long as we are outside.
B: Well, I only use sunscreen on sunny days. I don't use it on cloudy days.
A: That's not right. Apply sunscreen after your daily skin care routine, but before applying makeup. You'd better apply sunscreen 10 to 20 minutes before going out in order to form a protective film for the skin.
B: Do you have any good products to recommend?
A: It depends on how much time you spend outdoors. Generally speaking, the higher the SPF, the longer the protection, the more "+" after PA, the better the defense. Since you are taking children to the seaside, it is recommended that you use a physical sunscreen product, SPF 50, PA+++, waterproof and sweatproof.
B: What does that mean?
A: Well, physical sunscreen is safer, and less likely to cause sensitivity. It can even be used by pregnant women and babies.
B: Oh, so my baby can use it, too.
A: Yes.
B: Great.

Scanning Audio

1. Fill in the blanks with the words given below, changing the form if necessary.

| burden sunscreen apply waterproof protect physical pregnant spot |

(1) There are two types of sunscreen products, _____ sunscreen and chemical sunscreen.

(2) Summer is coming, you should use some _____ products to protect your skin.

(3) Please _____ the sunscreen 20 minutes before going out.

(4) I am _____. Can I use this cosmetic?

(5) This sunscreen is _____ and sweatproof. It is suitable for you to go to the beach.

2. Match the Chinese phrases with right English expressions.

防汗　　　　　waterproof

防晒指数　　　sweatproof

防水　　　　　SPF

斑点　　　　　spot

3. Arrange the following steps in correct order.

(1) glossing lip　(2) applying lip protector　(3) drawing lip liner　(4) applying lipstick

The correct order of lip makeup is: _____.

4. Translate the following sentences into English.

(1) 物理防晒产品给肌肤造成的负担更小。

(2) 每晚必须彻底清洁皮肤。

(3) SPF指数越高,防晒保护时间越长。PA"+"越多,防御效果越好。

(4) 孕妇和儿童适合用哪类防晒品?

(5) 防晒产品可以防止皮肤晒伤。

Critical Thinking and Group Peer-assessment

1. Change the customer types or demands, redesigning the conversation.
2. Each group tries to display the dialogue according to the text.
3. After one group finishes the dialogue, others make comments.

Knowledge Links

You Should Always Wear Sunscreen on Airplanes

We all know that we should be wearing sunscreen when we're in hot and sunny destinations. But did you know that you should be applying an SPF30 before you've even taken off for your holiday?

Dermatologists advise that we all wear SPF30 on a daily basis to prevent the signs of aging as well as skin cancer. But it turns out, it's no different when you're on board a plane. Though the windows on a plane are small, it's very bright when you're flying.

https://www.foxnews.com/travel/why-you-should-always-wear-sunscreen-on-airplanes

Unit 2
Body Care Products

Learning Objectives

1. Learn to recommend body care products in English according to their body characteristics and needs.
2. Be able to understand and translate the labels of body care products.

 Carrier Oil & Essential Oil

Both carrier oil and essential oil are the material basis of aromatic SPA. The fragrance of essential oil itself can directly affect people's physiological and psychological conditions, calm nerves and soothe emotions. Ms. Zhang has been under great pressure recently and suffered from insomnia. Her friend suggested she go to a beauty salon for an aromatic SPA.

Thinking and Talking

1. What is essential oil?
2. What are the differences between carrier oil and essential oil?

New Words

aromatic /ˌærəˈmætɪk/	a. 芳香的,芬芳的
	n. 芳香植物;芳香剂
aromatherapy /əˌrəʊməˈθerəpɪ/	n. 芳香疗法
physiological /ˌfɪzɪəˈlɒdʒɪkəl/	a. 生理的
individually /ˌɪndɪˈvɪdjʊəlɪ/	ad. 个别地,单独地
lavender /ˈlævəndə/	n. 薰衣草;淡紫色
jasmine /ˈdʒæzmɪn/	n. 茉莉
decompress /ˌdiːkəmˈpres/	v. 使减压,使解除压力
severe /sɪˈvɪə/	a. 剧烈的
dilute /daɪˈluːt/	v. 稀释;冲淡
concentrated /ˈkɒnsntreɪtɪd/	a. 集中的;浓缩的
gleam /gliːm/	v. 使闪烁;使发微光
insomnia /ɪnˈsɒmnɪə/	n. 失眠症,失眠
incense /ˈɪnsens/	n. 熏香
eucalyptus /ˌjuːkəˈlɪptəs/	n. 桉属植物,尤加利
dim /dɪm/	v. 使暗淡;变暗淡
penetration /ˌpenɪˈtreɪʃ(ə)n/	n. 渗透;突破;侵入

Phrases and Expressions

calm nerve	镇静安神
soothe emotion	安抚情绪
therapeutic effects	疗效
carrier oil	基础油

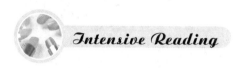

Intensive Reading

Text

Aromatic SPA is a kind of relaxation based on aromatic essential oil to promote metabolism through bath, massage, application of skin care products and aromatherapy. The fragrance of essential oil itself can directly affect people's physiological and psychological conditions, calm nerves and soothe emotions.

The essential oil can be used individually or mixed with many other kinds of essential oil to produce multiple therapeutic effects. For example, lavender and lemon can soothe and relax, and jasmine can decompress.

If applied to the skin directly, essential oils can cause reactions, such as severe irritation, redness or burning. Therefore, we usually use carrier oils to dilute concentrated essential oils, making them safe to use on the skin and helping "carry" them into the skin. In the meantime, carrier oils can also be used directly for facial and body massage to clear pores, smooth skin and gleam hair.

Conversation

(A beautician is welcoming her VIP customer, Ms. Zhang, at the door.)

A: Beautician B: Ms. Zhang

A: Welcome, Ms. Zhang.
B: I'm so stressed at work recently. I often have headaches during the day and insomnia at night. Is there a good way to solve my problems?
A: Oh! Then SPA care will surely help.
B: That's what I think.
A: This way, please, Ms. Zhang. This is our SPA room. Please come in. Please take a hot bath and relax. The water is ready for you. Would you please check the temperature?
B: OK, no problem.
A: I will light the incense for you. This is lavender incense. It can calm your mind. Lemon and eucalyptus essential oils have been added to the bath water to relieve your stress. I'll turn on the music and dim the light for you. Do you think that's all right?
B: Well, that's good! I feel very comfortable now.
A: You can soak for a while. Essential oil has small molecules and strong penetration power. Its efficacy is 60~70 times that of Chinese herbal medicine.

Therefore, it can directly dredge channels without massage, promote blood circulation and drive the whole body's qi and blood circulation.

(20 minutes later)

A: Let me dry your body and give you a massage with this carrier oil. It has an excellent moisturizing effect.

B: OK, thanks.

Practice

1. Fill in the blanks with the words given below, changing the form if necessary.

   > psychological individually insomnia decompress
   > essential aromatherapy massage directly

   (1) If you simply go to bed when you're sleepy and then get up at a fixed time, you'll cure your _____ and get a good skin.

   (2) If you need to _____ your pressure, you can choose SPA, shopping or listening to music.

   (3) It is difficult to heal _____ disease.

   (4) _____ oils are extracted from plants and are not harmful to humans.

   (5) It must be used _____, you should not mix it with others.

2. Match the Chinese phrases with right English expressions.

   薰衣草   lemon
   茉莉   eucalyptus
   柠檬   lavender
   尤加利   jasmine

3. Arrange the following steps in correct order.

   (1) massaging (2) taking a bath (3) moisturizing (4) exfoliating (5) masking
   (6) importing essence

   The correct order of aromatic SPA is: _____.

4. Translate the following sentences into English.

   (1) 轻音乐可以缓解紧张、放松心情。

   (2) 熏香是自我放松的方法之一。

   (3) 长期的精神亢奋会造成失眠。

   (4) 基础油可直接用于按摩。

   (5) 美美地睡一觉可以解除疲劳。

Critical Thinking and Group Peer-assessment

1. Change the customer types or demands, redesigning the conversation.
2. Each group tries to display the dialogue according to the text.
3. After one group finishes the dialogue, others make comments.

 ## Knowledge Links

What Is Geranium Essential Oil?

Geranium essential oil that has been commercially produced is a highly concentrated, volatile liquid extracted from the geranium plant typically through distillation or the use of chemical solvents. Natural beauty products, particularly those formulated to help alleviate conditions such as oily hair, often have geranium essential oil high on the list of ingredients. It might also be present in natural beauty creams or night creams that claim to help regulate oily or dry skin. Other benefits of geranium essential oil include its ability to help clear up acne, eliminate cellulite and kill lice, so it sometimes is added to hand or body lotions.

https://www.wisegeek.com/what-is-geranium-essential-oil.htm#didyouknowout

Task 2 Slimming Cream

 ## Lead-in

Most slimming creams on the market will have a burning sensation when used on our body. Sometimes plastic wrap is used to wrap up the place that needs to reduce weight. This kind of slimming cream has been gradually replaced by new products. Let's have a look.

Thinking and Talking

1. What is BMI?

2. Do you know any other ways to lose weight?

New Words

classification /ˌklæsɪfɪˈkeɪʃ(ə)n/	n. 分类；类别，等级
roll /rəʊl/	v. 卷；滚动，转动
elbow /ˈelbəʊ/	n. 肘部
armpit /ˈɑːmpɪt/	n. 腋窝
simultaneously /ˌsɪmlˈteɪnɪəslɪ/	ad. 同时地
convenient /kənˈviːnɪənt/	a. 方便的
previous /ˈpriːvɪəs/	a. 以前的；早先的；过早的

Phrases and Expressions

plastic wrap	保鲜膜
burning sensation	灼热感；烧灼感

Intensive Reading

Text

Product name: Slimming cream.

Use area: Any area that needs to be slimmed.

Classification: Massage cream.

Efficacy: Slimming, shaping, lifting and tightening.

Features:

1. Plant extract, mild and low sensitivity.

2. Direct rolling massage, easy to use.

3. Without plastic wrap, squeeze out the cream and massage directly.

Directions:

1. Double chin: pull from the bottom to the ear.

2. Arms: roll from elbow to armpit.

3. Legs: roll upwards.

4. Abdomen: massage in circles.

Notice:

1. It works best in the mornings and evenings or when the skin is relaxed.

2. Keep in one part for 10 minutes each time when you use it. Control your strength by yourself.
3. A few users with thin cuticle and sensitive skin may have red skin, which is a normal phenomenon.
4. No need to wash off immediately after use. It can produce continuous effects after massage and absorption.

Conversation

(A beauty consultant is introducing slimming products to Ms. Liu.)

A: Beauty consultant   B: Ms. Liu

A: Hello, Ms. Liu. Welcome.

B: Hi! I gained 3 kgs during the Spring Festival holidays at home. Look at my belly and my legs!

A: You can do a treatment of fat massage in our beauty salon, and push off excess fat in the stomach and legs. If you use the slimming cream simultaneously, the effect will be better.

B: Oh! What is it? Show me, please.

A: It's this slimming cream. It is convenient and safe. Previous weight-loss products have a burning sensation when applied to the skin, and the users may feel uncomfortable. This one won't hurt you. It's 100% pure plant based, very safe. In addition, there is no need for plastic wrap. It also has a ball for massage, which is very convenient.

B: How do I use it?

A: Massage the abdomen in circles and roll upwards when massaging the legs. It is better to stay in one area for 10 minutes each time when you use it.

B: OK. Should I wash it off immediately after the massage?

A: No, you'd better not. It produces continuous effects after massage and absorption.

B: Well, it sounds good. I'll buy one to try it out.

Scanning Audio

1. Fill in the blanks with the words given below, changing the form if necessary.

| convenience   slim   shape   lift   roll   elbow   extract   normal   share   effect |

(1) Proper exercise and proper diet can help you keep in good _____.
(2) You are welcome to visit our beauty salon frequently at your _____.

(3) Most people would forget about the skin on their _____, which is most likely to form cuticle.
(4) The main ingredients of this cosmetic are herb_____.
(5) Reducing calorie intake and cooperating with exercise can enhance the _____ of weight loss.

2. Match the Chinese phrases with right English expressions.

瘦身衣   slimming cream
减肥仪   exercise
瘦身膏   slimming clothes
运动   on a diet
控制饮食   weight reducing apparatus

3. Arrange the following steps in correct order.
(1) moisturizing (2) cleaning (3) shape analysis (4) weight reducing apparatus
(5) body mask (6) massaging (7) rinsing
The correct order of slimming care is: _____.

4. Translate the following sentences into English.
(1) 身材好的女孩能吸引更多的目光。
(2) 这款瘦身膏不需要使用保鲜膜。
(3) 它的成分是从植物中萃取出来的。
(4) 如果有更好的产品，我们会及时与您分享。
(5) 打圈式按摩腹部。

Critical Thinking and Group Peer-assessment

1. Change the customer types or demands, redesigning the conversation.
2. Each group tries to display the dialogue according to the text.
3. After one group finishes the dialogue, others make comments.

Knowledge Links

What Is BMI?

Body Mass Index (BMI) is a person's weight in kilograms divided by the square of height in meters. A high BMI can be an indicator of high body fatness and having a low BMI can be an indicator of having too low body fatness. BMI can be used as a screening tool but is not diagnostic of the body fatness or health of an individual.

If your BMI is less than 18.5, it falls within the underweight range.

If your BMI is 18.5 to 24.9, it falls within the normal or Healthy Weight range.

If your BMI is 25.0 to 29.9, it falls within the overweight range.

If your BMI is 30.0 or higher, it falls within the obese range.

https://www.cdc.gov/healthyweight/assessing/index.html

Task 3  Hair Removal Spray Foam

 Lead-in

Summer is the season that girls yearn for. They can wear various kinds of beautiful skirts and dresses. However, Ms. Sun has had excessive body hair since childhood. She is afraid to wear short-sleeved blouses and dresses in summer. She tried various methods to remove body hair, but the effect was not obvious and very painful. Do you have any good products to recommend to her?

Thinking and Talking

1. Does everyone need hair removal?
2. What hair removal methods do you know?

 New Words

foam /fəʊm/	n. 泡沫；水沫
depilate /ˈdepɪleɪt/	v. 脱毛
aloe /ˈæləʊ/	n. 芦荟
inhibit /ɪnˈhɪbɪt/	v. 抑制；禁止
allergic /əˈlɜːdʒɪk/	a. 对……过敏的

 Phrases and Expressions

hair removal   脱毛

tingling sensation	刺痛感
honey waxing	蜜蜡脱毛

Intensive Reading

Text

Product name: Hair removal spray foam.

Specification: 120 g.

Classification: Foam.

Main feature: Gently remove body hair.

Directions:

1. Apply a hot towel to the area where you want to depilate to soften the body hair.
2. Shake up and down before using.
3. Squeeze the foam to cover the part where hair removal is needed.
4. Keep it there for 10~15 minutes.
5. Wipe off the white foam with a paper tissue against the growth direction of the hair to get a smooth and white skin.

Notice:

1. The volume must cover the whole hair, too thin coating will weaken the depilation effect.
2. If your skin is comparatively thinner, there may be a tingling sensation. Apply aloe gel or skin care lotion after using the foam.
3. Please don't depilate continually.

Conversation

(Ms. Sun goes to a beauty salon to consult with a beauty consultant about depilation.)

A: Beauty consultant　B: Ms. Sun

A: Welcome, Ms. Sun. It's my pleasure to be at your service.

B: Summer is coming. Is there a good way to remove hair?

A: What methods have you used before?

B: I've tried everything. Hair remover! Hair removal cream! Honey waxing! But the effects are not satisfactory.

A: Honey waxing is very painful.

B: Yes! A lot!

A: I recommend you try this product. It is foam. It can remove your body hair gently.

B: Is it painful? How do I use it?

A: First, apply a hot towel to the area where you want to depilate to soften the body hair. Then squeeze the foam to cover the part where hair removal is needed. Remember to cover the whole hair, for too thin a coating will weaken the depilation effect. Wait for 15 minutes and then wipe off the white foam with a paper tissue against the direction of the hair. The natural ingredients can moisturize and smooth your skin. It is not painful at all. It also contains small molecules that inhibit hair growth.

B: Really? What about the effects?

A: You can have a try.

B: OK, I'll try it on.

Scanning Audio

1. Fill in the blanks with the words given below, changing the form if necessary.

   | foam   depilate   towel   waxing   squeeze   painful   cover   molecule |

   (1) The allergic reaction differs from individuals, some feels itching, some feels _____.

   (2) This hair removal cream can not only remove hair gently but also _____ hair growth.

   (3) You must _____ the whole hair with thick cream, otherwise, it will affect the depilation effect.

   (4) _____ is the most painful way to remove your body hair, even men are not brave enough to get that.

   (5) This hair removal can _____ your hair gently.

2. Match the Chinese phrases with right English expressions.

   蜜蜡脱毛　　　　　hair removal
   刺痛感　　　　　　inhibit
   脱毛　　　　　　　honey waxing
   抑制　　　　　　　tingling sensation

3. Arrange the following steps in correct order.
   (1) waiting for 10～15 minutes   (2) coating   (3) rinsing thoroughly   (4) body lotion   (5) wiping off with paper towel   (6) warming the skin
   The correct order of depilating is: _____.

4. Translate the following sentences into English.
   (1) 传统的脱毛方式是十分痛苦的。

（2）为什么我用了这款脱毛膏，皮肤有刺痛感？
（3）使用前请先上下摇动。
（4）毛巾应该定期清洗与更换。
（5）轻轻擦，不要用力。

Critical Thinking and Group Peer-assessment

1. Change the customer types or demands, redesigning the conversation.
2. Each group tries to display the dialogue according to the text.
3. After one group finishes the dialogue, others make comments.

Knowledge Links

Hair Removal Methods

- Shaving
- Bleaching
- Waxing
- Tweezing
- Lotions & Creams
- Threading

Caution: Do not remove hair if the areas have cuts, rashes, bumps, or sunburn. If you are going to swim or use sunscreen within 24 hours of removing hair, be careful. It can irritate your skin—so plan ahead!

http://www.pamf.org/preteen/mybody/girls/hairremoval.html

Unit 3
Hair Care Products

Learning Objectives

1. Learn to recommend shampoo, hair mask, hair dye and hair tonic to customers in English according to their hair types and needs.
2. Be able to understand and translate the labels of shampoo, hair mask, hair dye and hair tonic.

Task 1 Shampoo

Lead-in

Shampooing is the first step in the hair care routine. Shampoo has the function of cleansing hair and removing excess sebum from scalp. It is important to choose the right shampoo according to different hair and scalp types. As Ms. Li doesn't know what kind of shampoo suits her, the hairdresser recommends her to use anti-dandruff shampoo.

Thinking and Talking

1. What hair types are there?
2. What will you consider when you choose shampoo?

New Words

shampoo /ʃæm'puː/	n. 洗发水;洗发
	v. 用洗发露洗(发)
viscous /'vɪskəs/	a. 黏稠的;黏滞的
scalp /skælp/	n. 头皮
dislodge /dɪs'lɒdʒ/	v. (从既定和固定位置上)移开
dandruff /'dændrʌf/	n. 头皮屑
flake /fleɪk/	n. 薄片,一小片
worthwhile /wɜːθ'waɪl/	a. 值得的,值得花钱的

Phrases and Expressions

anti-dandruff	去屑的
hair type	发质类型
salon-crafted shampoo	沙龙级洗发水
daily shampoo	日用洗发水

Intensive Reading

Text

Shampoo is a kind of hair care product, typically in the form of viscous liquid.

Types of shampoo: Oily scalp shampoo, repairing shampoo and anti-dandruff shampoo.

Efficacy: Cleansing hair, removing excess sebum from scalp, dislodging dandruff flakes and keeping hair fresh.

Directions: Apply shampoo to wet hair and scalp, gently massage and rinse.

Conversation

(A hairdresser is recommending an anti-dandruff shampoo to Ms. Li.)

A: Hairdresser   B: Ms. Li

A: Good morning, Ms. Li. I'm glad to be at your service. Would you like to wash your hair?

B: Yes.

A: When choosing a shampoo, you need to consider your scalp and hair type. A healthy scalp is vital for healthy hair growth. But it seems that there is some dandruff on your scalp.

B: Yes, exactly.

A: Actually you will feel itchy with dandruff on your scalp, and in the long run, your scalp can be easily infected. So I suggest you buy an anti-dandruff shampoo in our salon, which can protect your scalp and reduce the damage from the environment.

B: That sounds good! What about the price?

A: As a salon-crafted shampoo, it's a bit more expensive than daily shampoo. After all, it contains milder ingredients and is more effective. It is really worthwhile.

B: May I try it this time?

A: OK. I'll be ready with it.

Practice

Scanning Audio

1. Fill in the blanks with the words given below, changing the form if necessary.

| worthwhile | shampoo | hairdresser | dandruff | inflamed | vital | excess | rinse | scalp |

(1) _____ is typically in the form of a viscous liquid.
(2) The _____ will recommend shampoo to you in the beauty salon.
(3) You should choose a shampoo that is suited for your _____.
(4) Oily scalp shampoo helps control embarrassing _____ oil and sebum production.
(5) This shampoo suits people who has _____.

2. Match the Chinese phrases with right English expressions.

油脂   root
受损发质   formula
头屑   damaged hair
配方   sebum
发根   dandruff

3. Arrange the following steps in correct order.

(1) wet hair with warm water   (2) comb hair before shampooing   (3) rinse thoroughly   (4) apply shampoo on hair   (5) gently massage hair and scalp

The correct order is:_____.
4. Translate the following sentences into English.
（1）您的头皮有点油,适合选用控油洗发露。
（2）这款洗发露可以温和清洁头发和头皮,清除头皮多余油脂,保持秀发轻盈。
（3）湿润头发后,涂抹少量洗发露并加以按摩,然后用水冲净。
（4）这款修护洗发水值得购买。
（5）专业洗发水的成分、功效与普通日用洗发水有所不同。

Critical Thinking and Group Peer-assessment

1. Change the customer types or demands, redesigning the conversation.
2. Each group tries to display the dialogue according to the text.
3. After one group finishes the dialogue, others make comments.

Knowledge Links

Before you choose shampoo, it's important to know what shampoo does, how to use it, and how that relates to your hair and scalp type.

Shampoo is meant to clean your hair and scalp of dirt, oil, and styling products. While you might think of shampoo strictly as a hair-cleansing agent, it's actually more important to consider your scalp when choosing a shampoo.

https://www.liveabout.com/shampoo-101-choosing-the-right-shampoo-3517815

Task 2 Hair Mask

Ms. Wang is a lady who has a love of beauty. She often perms or dyes her hair, as a result, her hair is badly damaged. She knows that she needs to have hair care regularly. However, she doesn't know the differences between conditioner and hair masks. She thinks using conditioner is enough. In fact, for damaged hair, it is necessary to use hair masks to give a deep conditioning regularly.

Thinking and Talking

1. Do you often use hair masks to nourish your hair?
2. How will you recommend hair masks to customers according to their hair types?

New Words

glycerin /ˈglɪsərɪn/	n.	甘油
frizz /frɪz/	n.	鬈发；鬈毛
restorative /rɪˈstɒrətɪv/	a.	恢复健康的
frizzy /ˈfrɪzɪ/	a.	（毛发）鬈曲的
unmanageable /ʌnˈmænɪdʒəbl/	a.	（头发）难梳理的
ventilated /ˈventɪleɪtɪd/	a.	通风的
perm /pɜːm/	n.	烫发
	vt.	烫（发）
residual /rɪˈzɪdjʊəl/	a.	剩余的；残留的
substance /ˈsʌbstəns/	n.	物质

Phrases and Expressions

hair mask	发膜
split ends	发梢的分叉
the acid and alkali value	酸碱值

Intensive Reading

Text

Product name: Hair mask.

Key ingredients: Water, glycerin.

Efficacy: Protecting hair from split ends and further damage, reducing hair frizz and leaving hair smooth and elastic.

Classification: Restorative hair masks for damaged hair;
Intense hydrating masks for dry hair;
Smoothing masks for frizzy and unmanageable hair.

Directions:

Apply to shampooed hair, comb through and massage. To enhance the super conditioning effect, wrap your hair in a warm towel. Leave it on hair for $10 \sim 20$ minutes. Rinse well.

Tip: Use hair masks once a week and less frequently as hair's health improves.

Notice: 1. Avoid contact with eyes. In case of contact, rinse immediately with water.

2. Store it in a cool, ventilated place and keep it out of reach of children.

Conversation

(In the beauty salon, the hairdresser is recommending Ms. Wang get a deep conditioning treatment for her hair after perming.)

A: Hairdresser B: Ms. Wang

A: You have just had your hair permed, Ms. Wang. Are you satisfied with the hairstyle?

B: I like it a lot, fits my style.

A: After perming, the hair becomes more fragile. I suggest you immediately give your hair a deep conditioning.

B: How does it work?

A: The pH of the treatment product is 4, which can quickly restore the acid and alkali value of the hair.

B: What if I do it a week later?

A: If you do it a week later, the residual alkaline substances can hurt your hair, and your hair may appear dry and dull, affecting the beauty of the hairstyle. So I advise you give your hair a deep conditioning. And we have a promotion this month.

B: What promotion?

A: The promotion is only for our members. You can choose to open a membership card. If you top up to 8,000 RMB, then you can enjoy a 50% discount.

B: OK, I will get an 8,000 RMB membership card today.

 *Practice*

Scanning Audio

1. Fill in the blanks with the words given below, changing the form if necessary.

| wrap hair mask membership elastic substance ventilated exclusively hydrate |

(1) _____ can be used after shampooing your hair.

(2) _____ a hot towel around your hair and leave it in place for 10 minutes, which will help optimize the mask's effect.

(3) Hair masks should be stored in a cool, _____ place.

(4) The hair mask is for damaged hair, which helps to restore _____ to hair.

  (5) This intense hydrating mask helps _____ hair.

2. Match the Chinese phrases with right English expressions.

  滋养　　　　　scalp treatment

  冲洗　　　　　nourish

  头皮护理　　　rinse

  弱酸的　　　　manageability

  易打理　　　　slightly acidic

3. Arrange the following steps in correct order.

  (1) heat the hair mask for 10~15 minutes　　(2) apply hair mask to towel-dried hair

  (3) shampoo hair　　(4) rinse

  The correct order is: _____.

4. Translate the following sentences into English.

  (1) 这款密集补水发膜适合干性发质。

  (2) 使用发膜可以预防发梢的分叉。

  (3) 建议一周使用一次发膜。

  (4) 这款护发产品的 pH 为 4,可以迅速恢复头发的酸碱值。

  (5) 充值 8,000 元可享受八折优惠。

Critical Thinking and Group Peer-assessment:

1. Change the customer types or demands, redesigning the conversation.
2. Each group tries to display the dialogue according to the text.
3. After one group finishes the dialogue, others make comments.

Knowledge Links

How to Take Care of Your Hair

  Maintaining your hair is relatively easy once you know how to care for it. Hair is made of protein, so keeping a healthy diet and practicing good hygiene are essential parts of maintaining luscious locks. If you want beautiful hair, start by washing and conditioning it properly. Then, learn the healthy ways to dry and style your hair. Finally, make healthy lifestyle changes to support healthy hair.

https://www.wikihow.com/Take-Care-of-Your-Hair

Part III Introduction of Cosmetics

 Hair Dye

 Lead-in

In order to look more fashionable, some young people choose to dye their hair into different colors. Only a few people are aware of the possible harmful health effects of hair dyeing. Some hair dyes contain chemicals which may increase a person's risk of cancer. If you need to dye your hair, you should follow some principles.

Thinking and Talking
1. What aspects do customers value most when they choose hair color products?
2. Do you have any suggestions for people who want to dye hair?

 New Words

permanence /ˈpɜːmənəns/	n.	持久性,永久
permanent /ˈpɜːmənənt/	a.	永久的
nontoxic /nʌnˈtɒksɪk/	a.	无毒的
expert /ˈekspɜːt/	n.	专家;行家
yellowish /ˈjeləʊɪʃ/	a.	微黄色的;发黄的

 Phrases and Expressions

hair dye	染发剂
semi-permanent hair dye	半永久性染发剂
plant-based hair dye	植物染发剂
patch test	局部测试
color chart	色卡
price list	价目表
light auburn	淡红棕色
allergy test	过敏测试

Text

Classification of hair dyes:

1. According to permanence of the effect:

   temporary hair dye, semi-permanent hair dye, permanent hair dye

2. According to the ingredients:

   Plant-based dye, chemical hair dye

Tips:

1. Choose milder, healthier plant-based hair dyes if you can.
2. Be sure to do a patch test for allergic reactions 48 hours before applying the hair dye. The hair dye can be used if the allergic reactions don't appear in 24 hours.
3. Don't dye your hair if there is a wound on your scalp, face or neck.
4. Don't shampoo your hair 24 hours before and after hair dyeing. Otherwise, the effect will be influenced.

Conversation

(A customer comes to the beauty salon to dye her hair.
The hairdresser is giving her some suggestions.)

A: Hairdresser   B: Ms. Wu

A: Have a seat please, Ms. Wu. Do you want to dye your hair today?

B: Yes. But I worry about the risk of hair dyes. It is said that most hair dyes contain chemicals, which can increase the risk of cancer.

A: I recommend you use a plant-based hair dye, which is nontoxic and safe to use. Experts also point out that there is no harm in dyeing hair twice a year.

B: Sounds good, I would like to try that.

A: What color do you want to dye your hair?

B: I have no idea. Do you have any suggestions?

A: Hair color should suit one's complexion. Your complexion is yellowish, so warm colors suit you more.

B: OK, I will go with your suggestion.

A: Very well. I'll get a color chart for you to choose. According to your current hair length, you can choose one from the following colors.

B: I will choose the light auburn.

A: Well, here are the brands and price list. Which brand of color cream do you prefer?

Part III Introduction of Cosmetics

B: I'll take this, my favorite brand.

A: All right. Some people may have allergic reactions to hair coloring, so we need to do an allergy test before dyeing your hair.

Scanning Audio

 *Practice*

1. Fill in the blanks with the words given below, changing the form if necessary.

| expert   complexion   plant-based hair dye   temporary   risk   harmful   price list   reaction   young |

(1) Long-term use of hair dyes may increase a person's _____ of cancer.
(2) After dyeing hair, some people look _____.
(3) Customers value the safety of hair dyes, they prefer _____.
(4) _____ suggest people not to dye hair more than twice in a year.
(5) If hair dyes are not used as directed, _____ health effects are possible.

2. Match the Chinese phrases with right English expressions.

   永久染发剂   classify
   分类   permanent hair dye
   预防,防备   introductions
   临时的   precaution
   说明   temporary

3. Arrange the following steps in correct order.
   (1) determine color (2) prepare to apply the hair cream (3) apply hair cream
   (4) communication before coloring hair (5) analyze customers' hair color and hair style (6) rinse and style the hair
   The correct order is: _____.

4. Translate the following sentences into English.
   (1) 染发剂有改变头发颜色的作用。
   (2) 我们应该尽量选用更温和、健康的植物型染发剂。
   (3) 染发前48小时做局部过敏试验,24小时无不适可正常使用。
   (4) 应该根据肤色选择适合自己的染发剂颜色。
   (5) 这是一款天然植物染发剂,具备抗褪色的功效。

 Critical Thinking and Group Peer-assessment

1. Change the customer types or demands, redesigning the conversation.

2. Each group tries to display the dialogue according to the text.
3. After one group finishes the dialogue, others make comments.

Knowledge Links

Even when hair dyes are used as directed, harmful health effects are possible. Up to 25 different ingredients in hair dyes can cause harmful skin effects. One of the main culprits is the primary intermediate PPD. Contact with skin can cause irritation including redness, sores, itching, and burning. Occasionally, allergic reactions occur and involve swelling of the face and neck that causes difficulty breathing. These toxic effects can occur immediately or up to a day after contact with the skin.

https://www.poison.org/articles/2016-sep/hair-dye

Task 4 Hair Tonic

Lead-in

Many women can suffer hair loss after delivery, which is called postpartum hair loss. Ms. Sun is one of them. She comes to the beauty salon to seek professional guidance. The hairdresser suggests her to try hair tonic.

Thinking and Talking

1. What kind of hair growth products will you recommend to your friends if they are suffering hair loss?
2. Can scalp massage help to stimulate hair regeneration?

New Words

mineral /ˈmɪnərəl/	n.	矿物
ethanol /ˈeθənɒl/	n.	乙醇
resist /rɪˈzɪst/	v.	抵制，阻挡

regeneration /rɪˌdʒenəˈreɪʃn/	n. 再生,重生
sparse /spɑːs/	a. 稀疏的;稀少的
greasy /ˈɡriːsɪ/	a. (头发或皮肤)油性的,多脂的
internally /ɪnˈtɜːnəlɪ/	ad. 内部地
delivery /dɪˈlɪvərɪ/	n. 分娩
cautious /ˈkɔːʃəs/	a. 谨慎的;小心的
follicle /ˈfɒlɪkl/	n. 毛囊

 Phrases and Expressions

hair tonic	生发液
drip tube	滴管
in moderation	适中;适量
postpartum alopecia	产后脱发
shampoo with ginger	生姜洗发水

 Intensive Reading

Text

Product name: Hair tonic.

Key ingredients: Mineral oil, ethanol.

Efficacy: Resisting hair losing and promoting hair regeneration.

Directions: Divide the hair into rows, and see the scalp as much as possible. Draw a small amount of hair tonic with a drip tube, drop in the hair loss or sparse area, and gently massage the scalp until fully absorbed.

Notice:

1. Hair tonic should be applied in moderation to prevent a greasy build-up in the hair.
2. Hair tonic can be poisonous, especially those containing ethanol, if taken internally.

Conversation

(A hairdresser is recommending hair tonic to Ms. Sun.)

A: Hairdresser   B: Ms. Sun

A: Your hair is thick, Ms. Sun. But the hair on the top is a little sparse.

B: Half a year after delivery, I'm still losing my hair.

A: You might have postpartum alopecia. You should be cautious of that.

B: Oh, I know that.

A: There is a shampoo with ginger. Followed with the hair tonic from the same series in our salon, the shampoo can control hair loss after pregnancy, stimulate the hair follicles, and promote hair growth through our professional massage.

B: Can I just buy these two products and use them at home? Will that be effective?

A: The application method of the hair tonic needs professional treatment. Only by combining with professional massage techniques to promote absorption, will the effect be better.

B: Then I'll come to your salon each time I need to apply the hair tonic.

Scanning Audio

Practice:

1. Fill in the blanks with the words given below, changing the form if necessary.

| hair tonic | hair loss | massage | absorb | promote | resist | scalp | follicle | delivery |

(1) _____ can stimulate hair regenerate and increase hair volume.

(2) As for her apparent _____, experts said it could have been stress-related.

(3) Apply a small amount of hair tonic on your scalp and _____ 3~5 minutes.

(4) Divide the hair into rows, and see the _____ as much as possible.

(5) A balanced diet and a healthy living habit can _____ hair growth.

2. Match the Chinese phrases with right English expressions.

    滴管   increase hair volume
    毛囊   hairline
    增加发量   stimulate
    刺激   hair follicle
    发际线   drip tube

3. Arrange the following steps in correct order.

   (1) massage the scalp   (2) clean your hands   (3) drip 2~3 drops of hair tonic
   (4) apply to areas where hair loss or hair thinning occurs
   The correct order is: _____.

4. Translate the following sentences into English.

   (1) 这款生发液可以防止脱发、促进毛发新生。
   (2) 用滴管汲取少量生发液,滴在脱发或稀疏区域,轻轻按摩头皮,直至被完全吸收。

(3) 您这属于产后脱发,要注意的。
(4) 生发液需要配合专业的按摩手法,可以促进吸收。
(5) 生发液的涂抹应该适量。

Critical Thinking and Group Peer-assessment

1. Change the customer types or demands, redesigning the conversation.
2. Each group tries to display the dialogue according to the text.
3. After one group finishes the dialogue, others make comments.

Knowledge Links

How Fast Does Hair Really Grow?

We are born with the total amount of hair follicles we will ever have over our lifetime. There may be about 5 million on our body, but our head has about 100,000 follicles. As we age, some follicles stop producing hair, which is how baldness or hair thinning occurs.

The American Academy of Dermatology says that hair grows about 1/2 inch per month on average. That's a grand total of about 6 inches per year for the hair on your head.

How fast your hair grows will depend on your:
- age
- specific hair type
- overall health
- other health conditions

https://www.healthline.com/health/beauty-skin-care/grow-hair-faster

Discussion Questions

1. What factors should we consider carefully when we buy cosmetics?
2. What is the effective way to control hair loss after pregnancy?

Part IV Practical Simulation Training

Task 1 Full Process Simulation Training for Medical Beauty Customer Service

Learning Objectives

1. Be able to clearly understand the basic service flow related to medical aesthetic service.
2. Master the use of English related to daily service of medical aesthetics to provide full service and follow-up to customers.

Lead-in

A customer needs to have a double eyelid operation, so she comes to a plastic surgery club. The staff of the club provides the whole service for the customer.

Thinking and Talking

1. How to make customers confident to take the operation?
2. How to carry out the work before and after the operation?

New Words

qualified /ˈkwɒlɪfaɪd/   a. 具备……的资历

aspirin /ˈæsprɪn/   n. 阿司匹林
anemia /əˈnɪmɪə/   n. 贫血
exaggerated /ɪɡˈzædʒəretɪd/   a. 夸张的

physical examination   体检
infectious disease   传染性疾病
cold compress   冷敷

 Intensive Reading

Text

(A customer comes to the beauty and plastic surgery institution to have a double eyelid operation.)

A: Beautician   B: Ms. Ma

A: Hello, welcome. Please come inside.

B: Hello.

A: Please take a seat here, and have some water first. May I have your name, please?

B: Thank you. My name is Ma Li. I made an appointment by phone yesterday.

A: OK, Ms. Ma, please wait a moment, and I'll confirm it. Yes, Ms. Ma, what kind of service would you like to do here?

B: I'd like to know something about the double eyelid surgery.

A: OK. Have you ever had plastic surgery before?

B: No.

A: All right, let's have a brief look first. The doctors and nurses in our institution are all qualified professionals with official certifications.

B: Do you have any requirements before the operation?

A: Well, please register here first, filling in your name, age, company and contact information.

B: OK, I'm done.

A: I'll fill in the remainder for you. Have you been smoking or drinking recently?

B: No.

A: Have you taken any drugs containing aspirin in the past two weeks?

B: No.

A: Are you menstruating or pregnant now?
B: No.
A: Since you are so young, I assume you don't have hypertension, heart disease, anemia, blood coagulation disorders, or any other diseases, right?
B: No. I have just finished a physical examination. I'm pretty healthy.
A: Do you have any infectious disease or other body inflammation?
B: No.
A: Well, let's go to the clinic and see the doctor.
B: OK.
(The beautician guides Ms. Ma to the doctor's office.)
A: Ms. Ma, this is Doctor Meng. He has many years of plastic surgery experience. He will provide you with a suitable double eyelid operation program.
B: OK, thank you. Doctor Meng, what kind of double eyelid would fit for my eyes?
Doctor: What are your ideal double eyelids in your mind, Ms. Ma?
B: I prefer them to be more natural, not too big, not exaggerated.
Doctor: Embedded double eyelid would meet your demand.
B: Is the recovery easy?
Doctor: There is no need to remove the suture after operation and the recovery period is short.
B: Sounds good. I'll take this.
Doctor: OK, my colleague will take you to pay the fee first, and then please wash your face, take photos and take a physical examination.
(Two days later, after confirming that all the physical examinations are normal, Ms. Ma enters the operating room, and has the operation under local anesthesia. After the operation is completed ...)
A: Ms. Ma, the operation has been completed. You may go home and have a good rest. Please pay attention to the following points:
  1. Ensure the surgical site is clean.
  2. Try to avoid water on the surgical site within 7 days after surgery.
  3. Avoid eating irritating foods such as peppers.
  4. Avoid smoking and alcohol.
  5. Strictly follow the doctor's instructions for medication and follow-up consultation.
  6. Pay attention to the exercise of eye muscles.
  7. Pay attention to the moisturizing and softening of eyelid tissue.
B: OK. How long will it take to recover?
A: The swelling of the eyelids will be obvious in 2~3 days and disappear in 5~6 days, and the ecchymosis will gradually turn yellowish. An ice bag can be used

within 24 hours after surgery to reduce swelling. Ten days after surgery, the swelling will gradually disappear and the double eyelid will be smooth, but not very natural. Two weeks to one month after the surgery, everything will return to normal, and the eyelids will be natural and elegant without any traces.

B: OK, I see.

A: If you need anything, please feel free to contact us. Please take care.

(On a follow-up call three days after the operation)

A: Hello, Ms. Ma. I'm Xiaoxue from the beauty and plastic surgery institution. I'd like to know about your current recovery from double eyelid surgery.

B: My eyelids are slightly swollen, and they are very uncomfortable.

A: Don't worry; it's normal.

B: OK. What else should I pay attention to?

A: You need to constantly do open-eye exercises so that you can recover more quickly.

B: Well, I will always ask you if I have any questions.

A: OK, I wish you an early recovery, and with becoming more and more beautiful. Goodbye.

B: Thank you. Goodbye.

 Practice

Scanning Audio

1. Fill in the blanks with the words given below, changing the form if necessary.

   | plastic   surgery period   exaggerated   physical examination |
   | recovery   avoid   swelling   certificate |

   (1) After the operation, you should _____ eating irritating foods.

   (2) You need to have a _____ before the operation.

   (3) The _____ organization is well known in the industry.

   (4) It is not suitable for plastic surgery during _____.

   (5) It will take at least three weeks for _____.

2. Match the Chinese phrases with right English expressions.

   双眼皮   infectious disease
   体检   pay attention to
   传染性疾病   double eyelid
   注意   physical examination
   专业的   professional

3. Arrange the following steps in correct order
   (1) greeting customers (2) physical examination (3) follow-up after operation
   (4) postoperative guidelines (5) registration of basic customers information
   The correct order is: _____.
4. Translate the following sentences into English.
   (1) 请问您之前有没有做整形手术的经历呢？
   (2) 我的眼睛适合做哪种类型的双眼皮手术呢？
   (3) 严格遵守医生嘱咐服药及复诊。
   (4) 您有任何需要随时和我们联系。
   (5) 我想了解一下您目前的双眼皮手术恢复情况。

Critical Thinking and Group Peer-assessment

1. Change the customer types or demands, redesigning the conversation.
2. Each group tries to display the dialogue according to the text.
3. After one group finishes the dialogue, others make comments.

Knowledge Links

Monster K-Cup Biggest Bra

Busty Brits can now buy the biggest bra ever seen on the high street—a colossal K-cup dubbed "the windsock". Until now the largest bra on general sale was a giant G-cup, and bigger ladies had to use specialist shops.

But department store Selfridges is introducing the enormous K-cup after a surge in demand from giant-boobed women. The bra has a whopping 4ft circumference and extra-thick straps to support its contents. Each cup measures an impressive 18ins at its widest point.

Britain's most popular bra size is a 36D, up from a 34B ten years ago. Manufacturer Fantasie says 10,000 women have already bought the K-cup-branded "the windsock" by fashion insiders—from specialist shops.

Selfridges' lingerie buying manager Helen Attwood said: "We are seeing more demand for larger sizes, especially fashionable bras for younger women."

The huge bra is too big even for fake-boobed glamour girl Jordan who was 32FF at her peak. But it still couldn't hold Britain's biggest breasts—a 40M pair owned by Donna Jones, 27, of Milton Keynes, Bucks.

Task 2 Full Process Simulation Training for Beauty Skin Care Customer Service

Learning Objectives

1. Be able to clearly understand the basic service processes involved in beauty skin care.
2. Master and use English related to beauty skin care to provide full service and follow-ups to customers.

Lead-in

A customer comes to a beauty salon and asks for skin care service. The staff of the salon provides full service to the customer.

Thinking and Talking

1. What are the common beauty skin care programs?
2. How to introduce nursing care to customers during the service process?

New Words

adjust /əˈdʒʌst/	v.	调整
lounge /laʊndʒ/	n.	休息室
consumption /kənˈsʌmpʃən/	n.	消费

Phrases and Expressions

turn over	转身
see sb. off	送别某人

Intensive Reading

Text

(A customer comes to a beauty club and asks for skin care services.)

A: Beautician   B: Ms. Fu

A: Good afternoon, Ms. Fu, welcome, please come inside.
B: Good afternoon.
A: Please have a cup of tea first. What kind of care are you planning to do today?
B: Back oil pressure and basic facials.
A: OK, Ms. Fu, please change your shoes here first.
   (Ms. Fu changes shoes)
A: Please follow me to the room.
B: OK.
A: Ms. Fu, please come in. You can put your belongings in the locker. This is a fresh bathrobe. We change it for each client, so please feel free to use it. While you change your clothes, I am going to prepare and will be back in five minutes.
   (After the preparation is finished, the beautician waits outside and taps the door.)
A: Ms. Fu, may I come in?
B: Please come in.
A: Is the room temperature still OK?
B: Fine.
A: Ms. Fu, it is three o'clock, and I am going to start now. The whole process will take about an hour and 45 minutes.
B: OK, let's get started.
A: Ms. Fu, since it's the first time for me to assist you, if there is any need to adjust the strength, please remind me at any time.
B: OK.
A: I will give you a back massage first. It can soothe and relax the nerves, waist, and back.
A: Now we have finished the back massage. How do you feel?
B: Good! It's comfortable!
A: OK, now please turn over and let's start doing the facial now. The first step is to remove the makeup; the second step is to clean the face; the third step is to tone the face; the fourth step is to exfoliate your skin; the fifth step is to apply massage cream; the sixth step is to apply a mask.
   (End of mask application)
The seventh step is to wash the mask away and apply skin-care products.

A: The whole program of care is over. Ms. Fu, are you satisfied with my service?
B: Yes, I feel much more comfortable.
A: Thank you! Ms. Fu, would you like to lie down for a while or get up?
B: I'd like to get up!
A: OK, then let me help you to get up. Please change your clothes and make sure that you have all your personal belongings with you.
B: OK, thank you!
A: Do you need to go to the bathroom? Or shall we go straight to the front desk to have some tea?
B: Let's go straight to the front desk. I'm a little thirsty.
A: OK, please come with me.
(They arrive at the reception lounge area)
A: Please have some tea first, and I will get your beauty registration card and record today's care information.
B: OK, please hurry up. I'm in a hurry.
A: Please confirm the service information. If it is correct, please sign here.
B: No problem.
A: When would you like to come back?
B: 2 p.m. next Monday.
A: OK, see you next Monday. I will see you off.
B: OK, see you next Monday.

 Practice

1. Fill in the blanks with the words given below, changing the form if necessary.

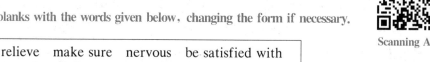

relieve   make sure   nervous   be satisfied with
apply   temperature   basic   consumption

Scanning Audio

(1) I would like to have a _____ facial care.
(2) You can _____ our mask three times a week.
(3) SPA is good for _____ our body.
(4) I _____ that Ms. Li doesn't have an appointment today.
(5) Customers _____ our service.

2. Match the Chinese phrases with right English expressions.

背部按压   facial cleansing
面部清洁   back pressure
敷面膜   service information
护肤品   apply mask

服务信息　　skin care products

3. Arrange the following steps in correct order.

(1) greeting customers　(2) nursing service　(3) ingredient preparation　(4) making reservation　(5) registration of customer care information

The correct order is: _____.

4. Translate the following sentences into English.

(1) 您今天打算做什么项目的护理呢?

(2) 整个过程大概需要 1 小时 45 分钟。

(3) 它可以舒缓、放松紧张的神经和你的腰、背。

(4) 您感觉怎么样?

(5) 请您换好自己的衣物并确保您的个人物品齐全。

Critical Thinking and Group Peer-assessment

1. Change the customer types or demands, redesigning the conversation.
2. Each group tries to display the dialogue according to the text.
3. After one group finishes the dialogue, others make comments.

Knowledge Links

Facial You Can Make at Home

Life is busy, a constant rush of responsibilities and appointments... and sometimes, you just need some time for yourself! If a day at the SPA seems impossible to fit into your busy schedule, grab some household ingredients and pamper yourself with a relaxing, DIY facial. Whatever your skin type, there's an easy-to-make option for you. Put on your favorite music, open the fridge, and get ready to mix and mash your way to the perfect facial.

Oily or Acne-Prone Skin: Egg White Facial

Egg whites tighten the pores and contain proteins that both repair skin and absorb excess oil, making them a powerful tool for dealing with acne. Separate the egg white from the yolk (save the yolk for cooking!). Whisk together the egg white with 1 teaspoon of lemon juice for one minute. The mixture will be messy, so apply it to your clean, dry face over a sink. Keep the mask on for 10 to 15 minutes — start whipping up your favorite dessert with that egg yolk while you wait! Once the mask starts cracking, wash away with warm water.

 Discussion Questions

1. What do you think of medical cosmetology?
2. Can we do facial care everyday?

参考文献

[1] 张卫华,余芊芊. 实用美容英语会话. 武汉:华中科技大学出版社,2017
[2] 张艳红. 美容美发实用英语. 北京:中国人民大学出版社,2014
[3] 杨玲. 美容美发英语(第二版). 北京:中国劳动社会保障出版社,2012

附录
课程标准

一、课程名称

美容实用英语。

二、适用专业及面向岗位

既适用于高职医学美容技术专业,又适用于中职美容美体、中医康复保健专业。面向美容导师、美容顾问、高级美容师、技术主管等岗位。

三、课程性质

本课程是医学美容技术专业的一门公共基础课程,其任务是培养学生具备美容行业英语基础知识,熟悉日常美容服务使用的基本用语,能够运用所学英语知识与客户进行交流,开展日常英语交流服务,具备美容导师、美容顾问、高级美容师、技术主管等岗位常用英语交流的基本能力。

四、课程设计

(一)设计思路

本课程标准的总体设计思路是:变学科课程体系为任务引领型课程体系,紧紧围绕完成美容机构工作人员日常工作任务所必须涉及的实用英语选择课程内容。变知识本位为能力本位,以岗位典型工作任务能力要求为依据,确定课程目标,以整体服务工作内容为导向设计学习单元,设定模拟学习情景,采用视频与图片贯穿教学的看、教、学、练一体的教学方式,培养学生运用英语知识于美容工作实践的交流会话能力。

(二)内容组织

本课程内容涵盖美容机构工作人员日常服务所涉及的英语基础知识,以突出实用、够用为原则,彻底打破传统学科体系的框架结构,基于工作过程典型工作任务的能力要求并结合学生认知规律,将学科系统知识整合为9个学习单元,每一单元下又划分为若干任务,任务内容与真实工作内容对接,学生学以致用,从而有效提高其学习主动性。

五、课程教学目标

(一)认知目标

(1)了解美容机构工作人员常用英语基础知识。
(2)熟悉用英语进行美容日常接待、产品介绍话术。
(3)熟悉美体服务中的英语交流话术。
(4)熟悉用英语进行美容服务全流程的介绍。

(二)能力目标

(1)针对美容美体及整形美容等各项目服务流程,正确地运用英语给客户做引导和介绍。
(2)根据美容产品的功能与适用人群,用英语为不同需求的顾客进行介绍。
(3)根据客户需求,正确地用英语进行介绍和说明。
(4)根据客户实际情况,用英语给出护理建议。

(三)情感目标

(1)具备用英语进行交流的专业素质。
(2)具备健康的体魄及良好的职业形象。
(3)具备良好的服务意识和团队合作意识。

六、参考学时与学分

参考学时:64学时;参考学分:3学分。

七、课程结构

序号	学习任务 (单元、模块)	对接典型 工作任务	知识、技能、态度要求	教学活动 设计	学时
1	接待客户	客户到店问候及店内引导	1. 了解在美容会所接待客户的流程，并能够用英语表述清楚 2. 用英语迎接客户进店 3. 用英语介绍店内的环境、设施及工作人员	1. 采用基于企业真实工作任务的案例教学法 2. 讲授结合小组讨论 3. 总结、归纳	6
		了解客户需求及登记注册	1. 了解如何用英语交流掌握客户基本情况及所需项目 2. 用英语交流分析客户基本情况并介绍护理方案 3. 用英语为客户做登记注册 4. 用英语指导客户交费 5. 用英语日常用语送客户离店并邀约其下次到店		
2	活动邀约及售后回访	电话邀约	1. 用英语在电话里熟练地介绍自己以及活动时间、优惠方案 2. 清楚地用英语介绍活动项目 3. 通过英语交流邀请客户到店参加活动	1. 理论知识应用于实际与客户交流对话 2. 案例教学：针对一名客户美容机构日常应当使用的礼貌用语	6
		售后回访	1. 用英语了解客户对会所的满意度 2. 用英语对顾客提出的质疑进行解释说明 3. 用英语鼓励客户继续坚持到店接受服务		
3	医美服务沟通	医美客户咨询	1. 学习使用医美客户咨询服务英语并向客户进行店内引导 2. 学会应用医美微整形手术所涉及的相关英语词汇 3. 学习用英语向医美客户介绍店内会员优惠活动	1. 基于如何用英语为医美客户服务的问题导入法 2. 基于用英语演示为医美客户服务流程的分组讨论	10
		术前沟通	1. 掌握术前英语沟通话术 2. 了解顾客病史及用药情况的英语沟通话术 3. 掌握医美手术的相关英语语句的翻译技能		

续 表

序号	学习任务 （单元、模块）	对接典型 工作任务	知识、技能、态度要求	教学活动 设计	学时
		术后注意提醒	1. 能够清楚地向客户用英语介绍医美微整形手术术后注意事项 2. 掌握医美微整形手术术后注意的主要英语词汇并加以应用 3. 学习用英语向客户介绍医美微整形手术术后饮食方面注意的话术 4. 学习用英语向客户介绍医美微整形手术术后运动方面应注意的话术	3. 基于如何用英语为客户做术前、术后及术后跟进的服务 4. 总结、归纳	
		术后跟进交流	1. 能够与医美微整形手术顾客用英语进行术后跟进的有效沟通 2. 熟练掌握及应用医美微整形手术术后跟进的专业英语词汇 3. 学会用英语安抚医美微整形手术术后顾客的焦虑心理		
4	美容皮肤护理交流与沟通	标准服务流程介绍	1. 能用英语表述皮肤护理服务的基本流程 2. 能用英语与客户在皮肤护理服务过程中进行交流 3. 掌握皮肤护理的相关英语词汇	1. 案例教学 2. 问题导入 3. 讨论、小结	8
		专业面部护理说明	1. 能用英语给客户分析面部皮肤检测 2. 用英语为顾客推荐面部护理方案，且英语表述清楚 3. 掌握皮肤种类和皮肤问题的相关英语词汇		
		面部服务流程沟通	1. 能用英语在给客户进行面部服务时进行交流 2. 掌握面部护理服务的基本流程，并能用英语表述清楚		
		家居护理建议	1. 能用英语表述家居皮肤护理的基本步骤 2. 能用英语给客户介绍家居皮肤护理并给出建议 3. 掌握家居皮肤护理的相关英语词汇		
5	美容美体保健	了解客户健康状况	1. 能用英语为客户讲解常用保健项目 2. 掌握美容会所咨询客户健康状况的常用英语表达 3. 能够根据客户健康状况用英语给出基本建议	1. 案例教学 2. 问题导入 3. 讨论、小结	8

续 表

序号	学习任务 (单元、模块)	对接典型 工作任务	知识、技能、态度要求	教学活动 设计	学时
		美容美体保健项目介绍	1. 了解如何用英语介绍美容美体项目 2. 掌握美容美体项目介绍的常用英语表达 3. 能够根据客户实际情况用英语为客户推荐具体项目		
		美容美体保健服务流程介绍	1. 了解如何用英语介绍美容美体保健服务流程 2. 掌握美容美体保健服务流程的相关英语词汇 3. 能够用英语与客户交流美容美体保健服务过程中的感受		
		保健注意事项提醒	1. 了解保健注意事项的常见英语专业词汇 2. 掌握美容美体保健注意事项的常用英语表达		
6	面部护理产品介绍	清洁类产品介绍	1. 掌握清洁类产品相关英文专业词汇 2. 能够看懂洁面产品英文标签 3. 能够用英文介绍洁面产品的作用和用法	1. 问题导入法：图片比较、用英语介绍面部护理产品 2. 案例教学 3. 小组讨论	6
		去角质类产品介绍	1. 掌握去角质类产品相关英文专业词汇 2. 能够看懂去角质类产品英文标签 3. 能够用英文介绍去角质类产品的作用和用法		
		补水类产品介绍	1. 掌握补水类产品相关英文专业词汇 2. 能够看懂补水类产品英文标签 3. 能够用英文介绍补水类产品的作用和用法		
		眼部护理产品介绍	1. 掌握眼部护肤品相关英文专业词汇 2. 能够看懂眼部护肤品英文标签 3. 能够用英文介绍眼部护肤品的作用和用法		
		抗皱类产品介绍	1. 掌握抗皱类产品相关英文专业词汇 2. 能够看懂抗皱类产品英文标签 3. 能够用英文介绍抗皱类产品的作用和用法		

续 表

序号	学习任务（单元、模块）	对接典型工作任务	知识、技能、态度要求	教学活动设计	学时
		防晒类产品介绍	1. 掌握防晒类产品相关英文专业词汇 2. 能够看懂防晒类产品英文标签 3. 能够用英文介绍防晒类产品的作用和用法		
7	身体护理产品介绍	精油类产品介绍	1. 掌握精油类产品相关英文专业词汇 2. 能够看懂精油类产品英文标签 3. 能够用英文介绍精油类产品的作用和用法	1. 问题导入法：用英语介绍身体护理产品 2. 案例教学 3. 小组讨论	4
		瘦身类产品介绍	1. 掌握瘦身类产品相关英文专业词汇 2. 能够看懂瘦身类产品英文标签 3. 能够用英文介绍瘦身类产品的用法		
		脱毛类产品介绍	1. 掌握脱毛类产品相关英文专业词汇 2. 能够看懂脱毛类产品英文标签 3. 能够用英文介绍脱毛类产品的用法		
8	养发护发产品介绍	洗发类产品介绍	1. 掌握洗发类产品相关英文专业词汇 2. 能够看懂洗发类产品英文标签 3. 能够用英文介绍洗发类产品的用法	1. 问题导入法：用英语介绍养发护发产品 2. 案例教学 3. 小组讨论	6
		护发类产品介绍	1. 掌握护发类产品相关英文专业词汇 2. 能够看懂护发类产品英文标签 3. 能够用英文介绍护发类产品的用法		
		染发类产品介绍	1. 掌握染发类产品相关英文专业词汇 2. 能够看懂染发类产品英文标签 3. 能够用英文介绍染发类产品的用法		
		生发类产品介绍	1. 掌握生发类产品相关英文专业词汇 2. 能够看懂生发类产品英文标签 3. 能够用英文介绍生发类产品的用法		

续 表

序号	学习任务 （单元、模块）	对接典型 工作任务	知识、技能、态度要求	教学活动 设计	学时
9	实战模拟训练	医美客户服务全流程介绍	1. 了解涉及医学美容服务的主要内容 2. 掌握涉及医学美容服务的基本流程 3. 运用涉及医学美容日常服务的英语，对客户进行全程服务与跟进	1. 问题导入法：用英语演示医美客户服务全过程及美容皮肤护理客户服务全过程 2. 案例教学 3. 小组讨论	8
		美容皮肤护理客户服务全过程介绍	1. 了解涉及美容皮肤护理的项目 2. 掌握涉及美容会所日常服务的基本流程 3. 运用涉及美容皮肤护理的英语对客户进行全程服务与跟进		
		机动			2
		合计			64

八、资源开发与利用

（一）教材编写与使用

本课程的总体设计是以美容技术岗位真实工作任务为载体，依据美容美体各项目的能力要求，将美容从业人员能够使用到的常用话术进行整理，突出岗位能力培养，因此，教材编写体现以下原则：

（1）彻底打破传统的各专业各学科一种英语课本、一种英语教学模式的"大锅饭"式的教学模式，本教材采取"自助餐"式的编写模式，选取美容从业人员所需的英语内容进行编排，不求量多，只求教学内容能够和美容实际工作相结合。

（2）突出实用、够用原则，必须具备高职特色，语法知识以够用为度，重点学习美容从业人员日常服务实用英语，使学生能够通过学习，顺畅地与客户进行交流。词汇及句法的分析不作重点，减少不用或少用的内容。

（3）体现以学生为中心的原则，编排内容图文并茂，课文对话简明、重点突出，方便自学，每一单元有学习目标，课后有练习题。

（4）在自编讲义的基础上，注重校本教材的开发和应用。编写的《美容实用英语》教材，体例有较大的突破和创新，图示清晰、醒目，文字简明。拟成为医学美容技术专业现代学徒制学生的专用教材。

（二）数字化资源开发与利用

充分利用校企资源平台，通过互联网教育平台建设，上传课件、视频、微课、知识链接、案例分析、习题等供学生复习、自学，形成丰富的课外教学互动资源。

积极开发和利用网络课程资源，建立课程在线考核评价系统，实现网上考核评价。

（三）企业岗位培养资源的开发与利用

开发新技术新项目，以真实案例为载体，开发项目培训课程。

九、教学建议

主讲教师具有丰富的英语教学经验，熟悉岗位工作任务及能力要求，积极推行工学结合、项目化教学，充分利用校企教育资源，加强与企业导师的合作与交流，使教学内容与实际工作任务紧密对接，学生学以致用。注重"教"与"学"的互动。

十、课程实施条件

由熟悉美容专业就业岗位及能力要求，有英语教学经验的教师授课，充分利用学校视频、图片资料资源等。注重教学内容与实际应用紧密对接，企业导师协助收集真实案例，作为教辅资料，增加课堂教学的趣味性。

十一、教学评价

（1）本课程的评价主要为结果考核及过程考核，考核形式采取笔试、面试、学习过程考核等。

（2）改革传统的学生成绩以理论考核为主、平时或实践成绩为辅的评价方法，采用一课一评、过程性评价与能力评价相结合，理论与实践相结合的形式，总评成绩以过程考核和操作能力考核占的比例较重（60%以上）。

（付明明）

"美容实用英语"课程内容结构

图书在版编目(CIP)数据

美容实用英语/付明明主编. —上海：复旦大学出版社，2019.6(2022.8重印)
全国现代学徒制医学美容技术专业"十三五"规划教材
ISBN 978-7-309-14322-5

Ⅰ.①美… Ⅱ.①付… Ⅲ.①美容-英语-高等学校-教材 Ⅳ.①TS974.1

中国版本图书馆 CIP 数据核字(2019)第 084234 号

美容实用英语
付明明 主编
责任编辑/陆俊杰

复旦大学出版社有限公司出版发行
上海市国权路 579 号 邮编：200433
网址：fupnet@fudanpress.com　http://www.fudanpress.com
门市零售：86-21-65102580　团体订购：86-21-65104505
出版部电话：86-21-65642845
上海四维数字图文有限公司

开本 787×1092 1/16 印张 9.25 字数 197 千
2019 年 6 月第 1 版
2022 年 8 月第 1 版第 2 次印刷
印数 4 101—6 200

ISBN 978-7-309-14322-5/T·646
定价：33.00 元

如有印装质量问题,请向复旦大学出版社有限公司出版部调换。
版权所有　侵权必究